IEE TELECOMMUNICATIONS SERIES 25

Series Editors: Professor J. E. Flood
                 Professor C. J. Hughes
                 Professor J. D. Parsons

PERSONAL & MOBILE RADIO SYSTEMS

Other volumes in this series:

PERSONAL & MOBILE RADIO SYSTEMS

EDITED BY
R.C.V. MACARIO

Peter Peregrinus Ltd. on behalf of the Institution of Electrical Engineers

Published by: Peter Peregrinus Ltd., London, United Kingdom

© 1991: Peter Peregrinus Ltd.

Peter Peregrinus Ltd.,
Michael Faraday House,
Six Hills Way, Stevenage,
Herts. SG1 2AY, United Kingdom

British Library Cataloguing in Publication Data
Personal and mobile radio systems.
  1. Great Britain. Radio services
  I. Macario, R. C. V.
  384.540941

ISBN 0 86341 219 X

Printed in England by Short Run Press Ltd., Exeter

Contents

Chapter 3

Modulation techniques: analog and digital

Chapter 4

Signal coding A: Channel coding

Signal coding B: Speech coding

Chapter 5

Conventional private mobile radio

Chapter 6

Trunked mobile radio systems

Chapter 7

Paging systems

Chapter 8

Switched networks in support of mobility

Chapter 9

The UK cellular system

Chapter 10

The Telepoint system

Chapter 11

Type approval for equipment used in mobile radio services

Chapter 12

Radio site engineering

Chapter 13

The pan-European cellular mobile radio system

Chapter 14

The pan-European cellular technology

Chapter 15

Future personal telecommunications

Preface

The contents of this book present enhanced versions of lectures given at the 1989 IEE Vacation School on personal and mobile radio systems. Because of the much greater and diversified interest in mobile radio since the Cellular Radio-telephone service was begun in January 1985, the technology of mobile radio has likewise progressed. This new text therefore has been designed to add to the original Peter Peregrinus book, *Land Mobile Radio Systems* published in 1985 and edited by R J Holbeche, and to complement *Data Communications and Networks 2*, edited by R L Brewster.

A week long vacation school contains a large number of lectures; however, experience has shown that in order to produce unified text it is really best to limit the number of contributions and a decision to this effect was taken early in the exercise. The contributions also tended to arrive serially and this explains why the appearance of this book appears somewhat delayed.

Nevertheless several advantages have been gained in the exercise. First of all the contributions have been cross-read and up-dated to the present *de-facto* European standards and use of words and definitions within the personal radio telephone business, made uniform where possible. Secondly several changes to the original layout and contents have been made possible. As a result the book is especially suitable for practising engineers and scientists both entering or who are in the personal radio business because the facts and definitions contained therein are not likely to change and will be a valuable source of reference for some time to come.

As editor I have taken the opportunity to discuss each contribution with the author or authors to encourage a uniformity throughout the text and a minimum of overlap or repetition. On the other hand no attempt has been made to 'steer' the text away from its basic educational and reference role. Personal and mobile radio systems in practice draw upon a vast wealth of theoretical concepts and advanced circuit technology. There is in fact a temptation to throw massive and diverse contributions into the book contents, but I believe the text would have become too weighty and possible lose its 'source of reference' status. Perhaps more encouragingly, it could point the way for further editions covering other matters which are becoming more closely specified, for example, the semiconductor technology needed for the realization of the various systems.

The list of contributors to this book also perhaps illustrates that we have tried to prepare a text evenly distributed across the many centres of research, development and operation involved in the personal communication business. Many other companies and colleagues have also had their advice sought on one topic or another, and I am very much indebted to many for such advice and their interest in this matter.

The overall preparation of the text really fell, however, to one person. Therefore I would like to record my special appreciation of the many hours of detailed work put in by Mrs Joy Knight to make every separate contribution appear exactly similar in layout. The time scale of the project has meant we have become quite skilled in layout; by self teaching! It would be unfair however not to mention the the considerable assistance also received from Rosalind Hann and Katie Petty-Saphon of the publishing department at Michael Faraday House. Truly the whole book is a joint effort.

R C V Macario
Swansea   January 1991

Chapter 1

Propagation models and observations

Raymond C V Macario

1.1 Introduction

Radio waves form part of the naturally occurring electromagnetic wave spectrum, any one frequency component of which can be represented as a progressive electromagnetic wave. Such a wave component is depicted in Fig. 1.1.

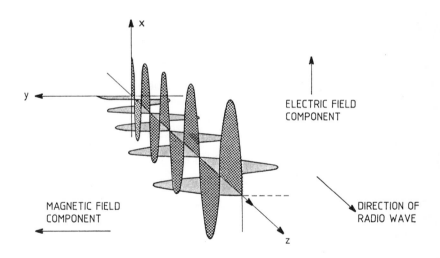

Fig. 1.1 Representation of an electromagnetic wave moving through space showing the magnetic and electric oscillatory components

This diagram is important to understand. Firstly the wave travels forward in direction and this is the basis of the concept of directivity of an antenna. The waveform repeats itself after a distance of its wavelength. For mobile radio this distance or wavelength ranges from meters to centimeters. Although the frequency of the wave, say 100 to 1000 MHz, is specified in the assignment of a channel to a mobile radio user, the wavelength is more relevant to the radiation and reception of the EM wave, when considering the antenna, and also to the defects of propagation of the wave, such as destructive multipath (directional) conditions.

Secondly Fig. 1.1 indicates that the EM wave has two active components, the electric field vector and the magnetic field vector, which are in phase in time and space. The diagram shows a vertically polarized radio wave. This polarization arose because of the way it was generated. Because there is a tendency to use vertically mounted dipoles or monopoles, vertical polarization ensues with a vertical electrical field component, usually measured in microvolts per meter or dBμV/m.

Unless one is able to markedly alter the electrical characteristics of free space, the magnetic component of the wave is closely related to the electric component value, from which one obtains the radiation power density (S) of the EM wave. This is measured as the electrical power (watts) passing through a plane one meter square, facing the direction of the wave.

The relation is given by the well known equation

$$S_R = \frac{E_R{}^2}{120\,\pi} \qquad \qquad ...(1.1)$$

120π is called the intrinsic impedance of the free space as can be seen from the relation between the voltage and power, i.e. approximately 377 ohms.

A third attribute of Fig. 1.1 is that if the page of the book is rotated the effect of a change of polarization can be noted. Ninety degrees gives a horizontally polarized wave; which will not be detected by a vertically orientated electrical antenna (dipole). It is to be noted that a change of polarization can occur at sharp angles of ground reflection.

A circularly polarized wave is one which has a continuous rotation of polarization induced into it by the radiating antenna structure. Current mobile radio systems use vertically polarized signals, but it may be found advantageous to use more complicated wave structures at new bands in excess of 1 GHz; indeed the signal is most likely to be very complicated as shown in the next chapter.

A final feature of Fig. 1.1 is that the diagram only shows a single EM carrier wave, i.e. of one frequency. An actual mobile radio signal will consist of a closely knit group of such waves, i.e. a modulated carrier wave. When discussing modulation the forward movement of the signal is not generally taken into account, but certainly does manifest itself under mobile conditions as described in Chapter 2.

For background reading on the more general aspects of an EM wave, books on Physics are recommended, e.g. Reference [1].

1.2   Research history

It will not be many years before it may be claimed that the propagation of radio waves has been researched for 100 years, depending on whether one begins with the UHF experiments of Hertz in 1887 or the LF experiments of Marconi about 1900. The big strides in the subject were made by T. L. Eckersley in the 1930's [2] who really understood the vagaries of radio wave propagation, how multipath arises and how signal dispersion can take place.

The writer was able to extend Eckersley's data to ELF and VLF in 1956 [3], whilst a valuable study on VHF and UHF wave propagation was published in 1965 by Matthews [4].

In the time scale being considered clearly many thousands of references on radio propagation have been published, and many studies are still in progress especially in regard to centimeter microwaves where new personal radio services are being considered. Only a few can be cited in this text, but for useful reading two compendium texts can be noted [5, 6].

One does not need to study all frequencies because mobile radio is allocated specific bands in the UK, at least. Table 1.1 has been adapted from Reference [7] and updated with the current regulatory information [8] which now supercedes the previously published UK Radio Frequency Allocations [9].

Table 1.1 is interesting in the fact that the trend is now towards higher frequencies and lower radiated powers. This fact will be found to alter our view on the propagation data required.

1.3   Antenna consideration

In order to describe radio wave propagation it is necessary to have a basic appreciation of antenna theory, but the requirements can be kept very brief. Thus the generation of the wave depicted in Fig. 1.1 required for example a conducting structure in which there is an oscillating electrical current.

The roof mounted monopole (half a dipole) is the best known example. As shown in Fig. 1.2 the current oscillates in amplitude and absolute magnitude along the vertical structure and sends out the radiation, shown in Fig. 1.1, mainly in the horizontal plane in which the antenna is situated. The fact that the radiation pattern is not isotropic implies directional gain (G). The gain and direction are related because antenna gain means a concentration of radiation in a particular direction.

A short dipole has a gain of about 1.5 times (= 1.76 dB) compared to an isotropic or point source antenna. Usually a dipole is physically half a wavelength long so as to present an ohmic impedance to the Tx or Rx of the system. Therefore at UHF, because the wavelength is now centimeters, the antenna which will be physically small both at the base station and at the mobile, and can be effectively regarded as a point source, when viewed at a distance.

On the other hand, when studying propagation and the receiving antenna it becomes helpful to think of an antenna as a collector of radiation, which has a certain aperture. This aperture is called the effective aperture (A) of the antenna and is the planar size of the antenna as far as a collector of radiation is concerned.

Band Name	Type of Service	Frequency band MHz	Maximum effective radiated power (ERP) watts
LOW All channels are at 12.5 kHz spacing	Mobile transmit Two-frequency simplex	71.50-72.80 76.95-78.0	25
	Base transmit Two-frequency	85.00-86.30 86.95-88.00	25
	Single-frequency simplex	86.3-86.7	5
MID/HIGH All channels are at 12.5 kHz spacing	Base transmit Two-frequency simplex	162.05-163.025 165.05-168.2375	25
	Mobile transmit Two-frequency simplex	157.45-160.5375 169.85-173.0375	25
	Single-frequency	158.5375-159.925	5
BAND III All channels are at 12.5 kHz spacing	Mobile transmit trunked duplex	192.50-199.50	20
	Base transmit trunked duplex	200.50-207.50	20
UHF Most channels at 12.5 kHz spacing	Base transmit Two-frequency simplex	440.0-446.0 453.0-454.0 456.0-457.0	25
	Mobile transmit Two-frequency simplex	425.0-429.0 459.5	25
UHF Cellular All Channels are at 25 kHz spacing	Base transmit Two-frequency duplex	917-933 and 935-950	10
TACS + ETACS	Mobile transit Two-frequency duplex	872-888 and 890-905	0.6/1.6/4
All systems below use digital modulation			
Cordless CT2	Duplex	864-868	0.1
Pan European Cellular or GSM	Base	950-960	required to operate with power control
	Mobile	905-915	
DSSR	Simplex	933-935	1
PCN's	Duplex	Band in 1710-1880	as yet unspecified

Notes: the ERP assigned to a service will depend on the range required and antenna siting and will often be less than the maximum value; also some VHF frequencies liable to change.

Table 1.1 - The frequency bands allocated to private and public mobile radio

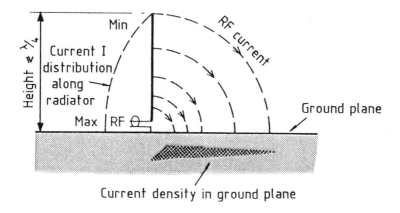

Current density in ground plane

Fig. 1.2 The RF current distribution for an active monopole antenna

An antenna of aperature A collects a power P (watts)

$$P = A.S_R \qquad ...(1.2)$$

It can be shown that the gain G and aperture A of all antenna are related by the formula [10]

$$G = \frac{4\pi}{\lambda^2} A \qquad ...(1.3)$$

The formula is most easily understood by considering a microwave dish antenna. The aperture is approximately the dish area (allowing for some loss of efficiency); therefore as the size (diameter) goes up, so does the gain.

For a short dipole,

$$A_{dipole} = G \frac{\lambda^2}{4\pi} = 1.5 \frac{\lambda^2}{4\pi} = \frac{3\lambda^2}{8\pi} \qquad ...(1.4)$$

Using equation 1.2 one can calculate the relation between the power received by a dipole P_d and the signal field strength (E) at the dipole. At 900 MHz a useful relationship is:

$$P_d \text{ (dBm)} = -135 - E \text{ (dB}\mu\text{V/m)} \qquad ...(1.5)$$

e.g. with 10 μV/m, P_d = -115 dBm i.e. higher than the usual receiver sensitivity. Lowering the frequency of operation to say 200 MHz, increases P_d to -103 dBm, because the aperture of a dipole tuned to 200 MHz is larger.

The signal power received is the critical factor in propagation because it has to overcome the noise power N_R at the receiver input. With N_R in dBm, the received signal/noise ratio when stated in decibels, will be

$$(S/N)_R = P_d - N_R$$

The situation can be improved by having a larger gain antenna. The result, with a receiving antenna of gain G_R, will be

$$(S/N)_R = P_R + G_R - N_R \text{ (dB)} \qquad ...(1.6)$$

For background reading on antennas many textbooks exist, e.g. Reference [10].

1.4   Model for propagation

Models for radio propagation all begin with the concept of two point source antennas in free space separated by a distance d (km).

The power density at the receiver of the radiation from the transmitter of source power P_T will be

$$S_R = P_T/4\pi d^2 \text{ (watts/m}^2\text{)}$$

since the radiation is distributed over a sphere of radius d. If the transmitter antenna has a directivity gain G_T, in the direction of the receiver the received signal increases to

$$S_R = G_T P_T/4\lambda \ d^2 \qquad ...(1.7)$$

Note $G_T P_T$ is the effective radiated power (ERP) of the transmitter.

If the receiving antenna has an aperture A, the power received as we earlier noted in equation 1.2, will be

$$P_R = S_R.A$$

Using equation 1.3, therefore

$$P_R = S_R.G_R \frac{\lambda^2}{4\chi\pi}$$

Introducing the value for S_R

$$\therefore P_R = (P_T/4\pi d^2) . G_T . (\lambda^2/4\pi) . G_R$$

or $\quad P_R/P_T = G_T.G_R (\lambda/4\pi d)^2 \qquad ...(1.8)$

Note that directivity gain at either end of the system enhances the received signal, but there remains the fundamental free space propagation factor $(4\pi d/\lambda)^2$ which reduces the received power as compared to the transmitted power. This is called the free space propagation loss factor L, and is given by:

$$L = (4\pi d/\lambda)^2 \qquad ...(1.9)$$

L arises due to dispersion of the wave energy. L increases as the wavelength reduces i.e. higher frequencies. This equation is more often written in logarithmic form:

$$L_{dB} = 32 + 20 \log f_{MHz} + 20 \log d_{km} \qquad ...(1.10)$$

Example 1   At 150 MHz over a path length of 1 km, the formula gives:
L = 76 dB

How does this relate to the transmitter power required to achieve effective communication? One needs knowledge of the receiver noise threshold N_R and the acceptable S/N ratio. Thus using equation 1.6

$$P_T > (S/N) + N_R + L - G_T - G_R \qquad ...(1.11)$$

G_T and G_R (now in dB) help the receiver as was also implied in equation 1.8.

Example 2

S/N	=	+20 dB
N_R	=	-120 dBm
L	=	76 dB
G_T, G_R	=	-3 dB (hand portables - helical antennas)

Therefore,

$$P_T > -18 \text{ dBm}$$
$$= 16 \text{ microwatts}$$

which is a very low power.

This is the free space result. On the ground specular reflection modifies the received signal (reflection coefficient of -1 assumed at VHF) and gives [4].

$$P_R = 4P_{direct} \cdot \sin^2 [2\pi h_T h_R / \lambda d] \qquad ...(1.12)$$

where h_T and h_R are the heights of the Tx and Rx above the plane earth. If one is working at ground level and the distance d > h (low angle of incidence), the equation simplifies to

$$P_R = 4P_{direct} \cdot [2\pi h_T h_R / \lambda d]^2$$

Introducing the modified P_R into equation 1.8 gives

$$P_R/P_T = G_T G_R.[h_T h_R/d^2]^2 \qquad ...(1.13)$$

Written in logarithmic form the propagation loss becomes

$$L_{db} = 40 \log d_m - 20 \log h_T h_R \qquad ...(1.14)$$

This equation is of fundamental importance to terrestrial mobile radio; it is an inverse fourth power law and is independent of frequency. It is also the basis of CCIR TV and Radio coverage data [11]. Note the distance d is now in metres, not kilometres as in equation 1.10.

Example 3

    Same distance d = 1000 m = 1 km
    h_T x h_R = 10 m² (low antennas)

Therefore   L   = 100 dB

P_T required now is 4 milliwatts for a flat earth 1km radio cell.

Example 4

    Distance d   = 25km
    h_T x h_R = 100m² (base station set high)

Therefore   L   = 136 dB

P_T required is now 16 watts; hence Table 1.1.

Observations [5, 6, 12] show that, apart from the effect of the earth's curvature over a long path (large radio cell), several terrestrial effects must be taken into account, which include

    (i)   Surface roughness
    (ii)   Line of sight obstacles
    (iii) Buildings and trees, etc.

The propagation loss equation is therefore amended to read

$$L = 40 \log d - 20 \log h_T h_R + ß \qquad \text{...(1.15)}$$

where ß are the additional losses, lumped together, which are studied further below. If the additional loss factor ß is a constant this equation says the radio cell (range equal to the radius) is circular. Because ß, and h_R can vary according to which direction one views from the transmitter, most radio cells are not circular.

1.4.1 Earth curvature effect

If one plots the path profile in any direction (radially) from the transmitter the horizon will soon vanish because of the curvature of the earth, i.e. Fig. 1.3. The radio horizon however extends further, due to the variable refractive index of the atmosphere [4, 6]. It is generally agreed that the bending adjustment or K-factor is about 4/3. The path profiles for radio propagation are therefore usually plotted on 4/3 profile paper, an example of which is shown in Fig. 1.4.

This is a path going north from South Wales over a mountainous region. The 'line-of-sight' path A is for a microwave link system and eqn 1.10 would apply. Path B on the other hand is a path where there is a knife edge obstacle, between the transmitter and a receiver situated in the shadow of a particular mountain peak.

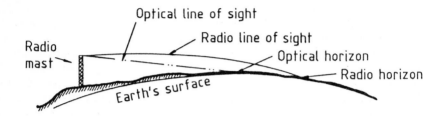

Fig. 1.3   Optical line-of-sight compared to a radio line-of- sight horizon

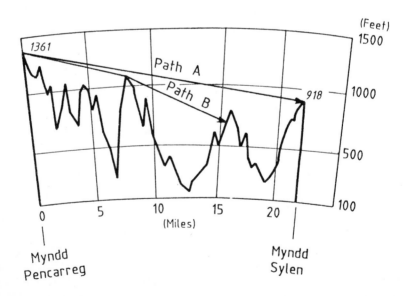

Fig. 1.4   4/3 path profile terrain plot across a mountainous region of Wales, showing two possible paths

1.4.2 Knife edge diffraction

Because the radio wave Fig 1.1, is not a line, but extends either side of the ray path, diffraction at shadow edges occurs, such as for path B in Fig. 1.4. Energy appears in the shadow, or the Fresnel zones as they are called, which can be given as an attenuation below the direct ray. Fig. 1.5 indicates the type of result that can be expected [4, 13]. The information is frequency dependent because the Fresnel zones become smaller with increasing frequency.

The position of the knife edge is at the point zero on the horizontal scale. In plane earth theory half the signal would become completely shadowed. For a sharp 'knife edge', such as the contour in Fig. 1.4 a 6dB attenuation would result if the path just cleared the peak. In the shadow of the peak the attenuation gradually increases, but some signal is observed due to the wavefront spreading of the RF wave. The same spreading also causes amplitude fluctuation of the signal when the line of sight clears the obstacle, i.e. path A in Fig. 1.4.

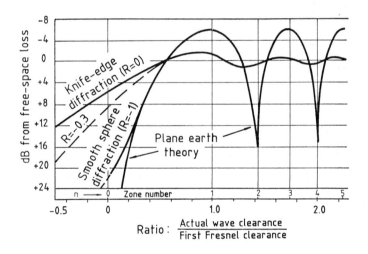

Fig. 1.5   Additional attenuation of a direct ray when
in the vicinity of an obstacle

1.4.3 Rough ground model

Rough ground causes interference between the direct ray and reflected rays. Fig. 1.6 [5] shows the type of rough grounds which may be encountered. Clearly the calculation of the received signal, even at one distance along one radial is a complicated procedure.

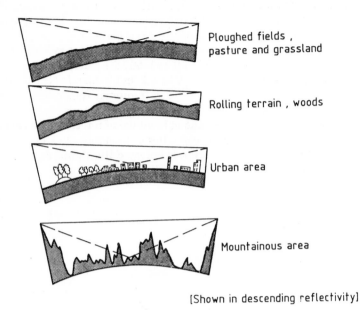

Ploughed fields , pasture and grassland

Rolling terrain , woods

Urban area

Mountainous area

[Shown in descending reflectivity]

Fig. 1.6   Types of 'rough ground' which can be encountered
by a radio wave   (Adapted from Ref. 5)

1.5   Empirical models

The basic formula, eqn 1.15 is not complicated except we do not know β.   Also it appears to be independent of frequency.   This is certainly not the case.   All observations show the radio coverage decreases with increasing frequency.   A clear example is shown in Fig. 1.7 [12];   going from 168 MHz to 900 MHz one looses 20 dB.

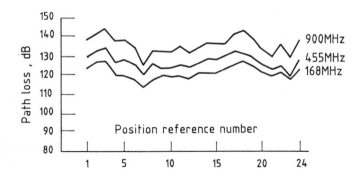

Fig 1.7   The influence of frequency on the excess propagation
loss term from observation   (From Ref. 12)

Empirical model formula are based on force fitting formula to measured data. A good example is the formula for British Cities by Allesbrook and Parsons [12]. They proposed an equation of the form

$$L_{dB} = 40 \log d_m - 20 \log h_T h_R + 20 + f/40 + 1.08 L - 0.34H \qquad ...(1.16)$$

Where   f = frequency in MHz
          L = land usage factor - the percentage of the test area covered by
by   buildings of any type, 0 - 100%
          H = terrain height differences between the T_x and R_x, i.e.
              (R_x terrain height - T_x terrain height)

Example 5

    f = 160 MHz
    L = 30%
    H = 50 m (over the hill)

Therefore, excess loss, β = 20 + 4 + 6 + 15
                            = 45 dB

This additional 45 dB loss implies that a 125 W base station is now required to cover a path distance, or cell size, of 25 km.

An additional loss factor based on an urbanisation factor U is introduced in the model for use in highly urbanised city centres [14]. Data was only available to calculate U, for test area covered by buildings of more than four storeys, such as central London. This additional factor may normally be neglected, except in such locations.

Ibrahim and Parsons derived their model with the aid of measurements taken. Actual received signal strength was compared to that predicted by the plane earth equation alone, and the difference (excess clutter factor) found. Their final model is therefore composed of the plane earth equation plus a clutter factor which is a best fit equation based on the factors considered most likely to increase propagation loss.

The above model is perhaps pessimistic for low antenna heights < 10 m within small cells, and does not include a through building loss factor, B. A model better suited to micro cell applications is:

$$L_{db} = 40 \log d_m - 20 \log h_T h_R + 20P + (f - 145)/40 + 0.1L + U + B \quad ...(1.17)$$

The factor, P, varies between zero and one, and depends on the two antenna heights. Assuming one antenna, < 3 m, then the other antenna height > 20 m, P = 1, but for antenna heights < 2 m, P => 0.

U is an excess loss factor which may be included to account for the urbanisation factor plus any other factors giving rise to significant deviations.

Observations of the factor B, the building loss factor, by various studies indicate that $0 < B < 32$ dB; i.e. References [15, 16, and 17].

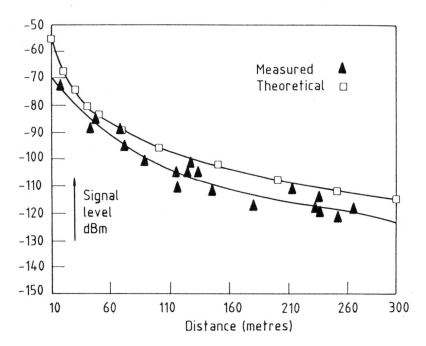

Fig. 1.8   Measured signal strength from mobile station in a
building to a portable station in an urbanized area

Path loss data is shown in Fig. 1.8 for a typical city conurbation site in one direction. The same result is unlikely in another direction because of the nature of the infrastructure. Using a Tx power of one milliwatt (0 dBm) and a 0dB antenna gain factor, the path loss ordinate is then equal to the received signal strength in dBm (with the sign changed). Ranges of 100 m to 200 m for a -120 dBm receiver threshold sensitivity are therefore expected, and indeed observed, i.e. this would be the typical cell size.

The small cell, in an urban area result, suggests a simple model

Propagation Equation

LOSS   = Open Space   +   Building +   In/out
in dBs   loss   clutter   building

           = L_o +   (30 - 40) +   (20 - 30)
                                   dB   dB

This relation actually suggests a simple rule of thumb for personal radio propagation, namely:

1 microwatt becomes 1 μV/m before you start.

The IEEE Vehicular Technology Society recently reviewed propagation models [18] at 900 MHz. The following formula are of interest:

40 log d - 20 log h + 139   (based on measurements by Egli)
55 log d - 19 log h + 111   (based on measurements by Carey)
37 log d - 20 log h - 142   (based on measurements by Lee)

The problem with the empirical models, as we have just seen is that they all have various adjustment factors depending on the location, different mobile heights, etc., but they are useful for initial proposal exercises.

The three formula are plotted in Fig. 1.9. There is really not too much difference in the end result, except for two important factors.

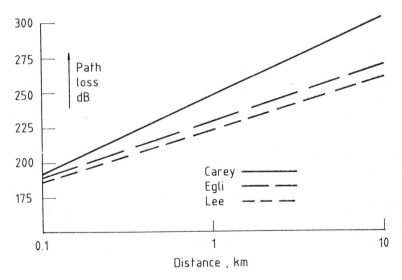

Fig. 1.9   Propagation loss prediction based on three North American models

The differences suggest the inverse fourth power distance result is not absolute. Close to a transmitter, such as in a microcell, some measurements suggest the inverse distance (free space) law operates masked by the clutter factor ß.

To assume greater than a fourth power distance law operates is however optimistic with respect to co-channel interference problems which arise when designing a cellular radio system, e.g. Chapter 9.

It is perhaps also of interest to draw attention to the variation of signal with distance as one goes from the very lowest frequency radio signals to microwaves. At VLF the signals follow an inverse distance law, which gradually meets increasing attentuation as one goes to LF and MF frequencies. Within the HF band an inverse distance law again almost applies, coupled to extensive multipath modes. At VHF, as just discussed, an inverse fourth power law applies which then gradually meets increasing attenuation as one goes up to UHF frequencies. At the higher UHF, or low microwave frequencies, specular short distance reflection causes extensive multipath and also the propagation law at short distances almost returns to the inverse square of the distance law.

Finally, the drawback, as with indeed all other of these formula, is they more or less predict circular radio cells, which is rarely the case, at least in the UK.

1.6   Computer prediction programs

A reasonable estimate of the coverage area of a transmitter can be achieved by two means; (i) using a repeater technique will soon provide an estimation of the radio coverage (cell boundaries) and also does not require the transmitter to be continuously operating; (ii) use a field strength measuring receiver attached to a distance or location monitor. New instruments, hand portable or vehicle mounted can achieve very impressive field strength maps.

For a detailed theoretical predicted field strength coverage map a precise topological map of the surrounding countryside is neccessary. In the UK such a data base was first described by Edwards and Durkin [19] and this program has been gradually developed by the Joint Radio Committee (JRC) of the Supply Industries into a general radio coverage prediction program [20].

Similarly, since the topology of the UK is now available as a data base from the Ordnance Survey Services in 50 meter squares over the country and in towns both the BBC and the IBA have extensive propagation prediction and frequency planning services, along with the two 900 MHz Cellular Radio operators in the UK and other industries.

The programs not only plot out the expected service area for a particular site, but more importantly predict where co-channel interference can be expected in shared trunked radio cells, for example, indeed at well displaced areas of the UK Access to these programs is possible by negotiation. The programs are based on the calculation procedures outlined above, and experienced force-fitting of the data to existing empirical results.

1.7   Summary of propagation conditions

The CCIR data curves [11] for TV and Radio are almost independent of frequency. Also the receiver signal levels expected do not assume marginal conditions of reception. This is not so for mobile radio. Poor signal operation is expected and limited transmitter ERP is mandatory. Frequency now enters in the propagation data and limits range at higher frequencies. At lower frequencies man-made electrical noise raises the receiver noise floor. The general scene can be illustrated by Table 1.2. Undoubtedly the UHF television bands are optimum for mobile radio, yet it is most unlikely any of these frequencies will be released for LMR [8].

Attribute	VHF Bands MHz			UHF Bands MHz		
	60	80	160	450	900	1800
Communication* distance-town, km	12	13	14	9	4	1
Communication* distance-country, km	20	17	14	9	6	3
Monopole antenna height, m	3	2	1	0.3	15cm	7cm
Building penetration	poor	poor	mod	good	good	good
Multipath fading rate	slow	mod	mod	fast	fast+	serious
Effect of vehicle and other noise	serious	poor	mod	small	receiver noise figure decides	

*The distance here considers the performance of an average sited base station communicating with a vehicle

Table 1.2 The propagation attributes of the main private mobile radio bands

One factor missing from Table 1.2 is any referral to anomalous propagation. Three cases may be cited:

1   Band I: Sporadic-E causes specular reflection by clusters in the iono-sphere and gives rise to signal hopping in the range 800 - 2600 km [21]. This can especially cause TV interference to intended mobile services [22] in the shared band.

2.   Band III, etc: Ducting in atmosphere, especially during hot weather, means normal signals can get directed away from the ground [23].

3.   New 1.8 GHz band: The small wavelength (16 cm) could give rise to more complicated excess loss factor due to more apparent topological details, i.e. road signs. The propagation loss at these frequencies is generally greater than at half the frequency, i.e. 900 MHz [24].

The signal level we have been concerned with is what is known as the *mean signal level* in a location, independent of any multipath and moving effects; the above phenomena affect the mean level, on top of which multipath effects must be considered as described in the next Chapter.

1.8   References

1.   SOLYMAR, L.: 'Lectures on electromagnetic theory' (Oxford Press, 1984)

2.   ECKERSLEY, T.L.: 'Studies in radio transmissions', Proc. IEE, 1932, 71, p. 405

3.   CHAPMAN, F.W., and MACARIO, R.C.V.: 'Propagation of audio frequency radio waves to great distances', Nature, May 1956, 177, p.930

4.   MATTHEWS, P.A.: 'Radio wave propagation, vhf and above' (Chapman & Hall, London, 1965)

5   SHIBUYA, S.: 'A basic atlas of radio propagation' (Wiley, New York, 1983)

6.   FREEMAN, R.L.: 'Radio system design for telecommunications 1-100 GHz' (Wiley, New York, 1987)

7.   MACARIO, R.C.V.: 'Mobile radio telephones in the UK' (Glentop, UK, 1988)

8.   DTI RD.: 'Report of the civil spectrum review committee: Stage 1: 470-3400 MHz', London, April, 1989

9.   DTI RD.: 'UK table of radio frequency allocations' (HMSO Publication, 1987)

10.   GRIFFITHS, J.: 'Radio wave propagation and antennas' (Prentice-Hall International, 1987)

11.   EBU Report: 'Planning parameters and methods for television broadcasting in the VHF/UHF bands', Report No. 3254, May 1988. Data based on CCIR recommendation 370

12.   ALLESBROOK, K., and PARSONS, J.D.: 'Mobile radio propagation in British cities at frequencies in the VHF and UHF bands', Proc. IEE, 1977, 124(2), p. 95

13.   PARSONS, J.D., and GARDINER, J.G.: 'Mobile communication systems' (Blackie, London, 1989)

14.   IBRAHIM, M. F., and PARSONS, J.D.: 'Signal strength prediction in built-up areas, Part 1', Proc. IEE, 1983, 130, Part F, (8), p. 377

15.   RICE, L.: 'Radio transmissions into buildings at 35 and 150 mc', Bell System Technical, 1959, 38, Jan. p.197

16.   HOFFMAN, H., and COX, D. : 'Attenuation of 900 MHz radio waves propagation into metal buildings', Trans. IEEE AP-30, 1982, July, p.4

17.   TURKMANI, A.M.D., PARSONS, J.D., and LEWIS, D.G.: 'Measurement of building penetration loss on radio signals at 441, 900 and 1400 MHz', J.I.E.R.E.,1988, 58, S6, December (Supplement), p. 169

18.   IEEE Vehicular Technology Society, 'Special issue on mobile radio propagation', 1988, 37, February

19.   EDWARDS, R., and DURKIN, J.: 'Computer prediction of service area for VHF mobile radio networks', Proc. IEE, 1969, 116, No. 9, September, p.1493

20.   Joint radio committee for National services, Millbank, London

21.   EDWARDS, K.J., KERSLEY, Y., and SHUBSOLE, L.F.: 'Sporadic-E propagation at frequencies around 70 MHz', J.I.E.R.E., 1984, May, p.54

22.   MACARIO, R.C.V., and GRIMM, F.: 'An investigative study of services which could be implemented in UK Band I radio spectrum', IERE Land Mobile Conf, 1985, Cambridge, p.1

23.   TOWNSEND, A.A.R.: 'Analog line-of-sight links' (Prentice Hall International, 1987)

24.   PARSONS, J.D.: 'Propagation characteristics in mobile radio frequency bands', Telecommunications J., 1989, December, p. 50

Chapter 2

Characterisation of fading mobile radio channels

Professor David Parsons

2.1 Introduction

Radio communication links are often subjected to conditions in which energy can travel from the transmitter to the receiver via more than one path. This "multipath" situation arises in different ways depending upon the application; in mobile radio it is caused by reflection and scattering from buildings, trees and other obstacles along the path. Radio waves therefore arrive at a mobile receiver from different directions, with different time delays. They combine vectorially at the receiver antenna to give a resultant signal which depends crucially on the differences in path length that exist in the multipath field, since the various spectral components that comprise the transmitted signal can combine constructively or destructively. As a vehicle-borne, or hand-held, receiver moves from one location to another the phase relationship between the spectral components within the various incoming waves changes, so the resultant signal changes rapidly. It is worth noting that whenever relative motion exists there is also a Doppler shift in the frequency components within the received signal.

Characterising the mobile radio channel is not a simple task. It is possible, however, to deal with the problem on two fronts. Firstly we can provide a characterisation that is adequate for the case when the signals occupy only a narrow bandwidth. By "narrowband" in this context we mean that the spread of time delays in the multipath environment is sufficiently small so that all spectral components within the transmitted message are affected in similar way. In other words, there are no "frequency-selective" effects and the characteristics of the channel can be expressed in terms of their effect on any one component in the message - the carrier frequency is commonly used. A more complicated form of characterisation is needed to deal with wideband signals, "wideband" in this case being used to indicate that frequency-selective effects do occur.

First we deal with the narrowband case in order to introduce some of the ideas and terminology relevant to this subject. In urban areas problems exist due to the fact that the mobile antenna is low (so there is no line-of-sight path to the base station) and is often located in close proximity to buildings. Propagation is therefore mainly by means of scattering and multiple reflections from the surrounding obstacles as shown in Fig. 2.1. Because the wavelengths in the VHF and UHF bands are less than 1 metre, the position of the antenna does not have to be changed very much to change the signal level by several tens of dB. The rapidly changing signal is basically spatial phenomenon, but a receiver moving continuously through the field experiences a time-related variable signal which is further complicated by the existence of Doppler shift. The signal fluctuations are known as fading and the rapid fluctuations caused by the local multipath are known as fast fading to distinguish them from the much longer term variation in mean level which is known as slow fading. This latter effect is caused by movement over distances large enough to produce gross variations in the overall path between the base station and the mobile. The typical experimental record shown in Fig. 2.2 illustrates these characteristics. Fades of a depth less than 20 dB are frequent but deeper fades in excess of 30 dB, although less frequent, are not uncommon. Rapid fading is usually observed over distances of about half a wavelength and at VHF and UHF, a vehicle moving at 50km/hr can pass through several fades in a second. If we consider the nature of the multipath and the shadowing effects, and their influence on the characteristics of radio wave propagation, it is apparent that it is pointless to pursue an exact or deterministic characterisation and we must resort to statistical communication theory.

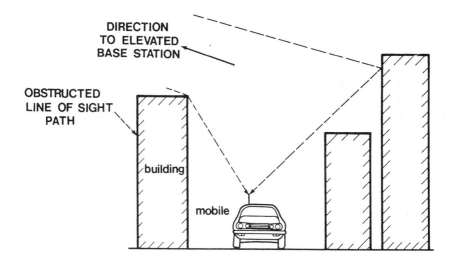

Fig. 2.1   Illustrating the mechanism of radio propagation
in urban areas

Examination of Fig. 2.2 shows that it is possible to draw a distinction between the short-term multipath effects and the longer-term variations of the local mean. Here, however, we consider only the short-term effects both for narrowband and wideband channels, in other words, we consider the signal statistics assuming the mean value to be constant.

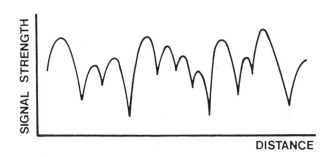

Fig. 2.2 Typical fading envelope in an urban area

2.2 The nature of multipath propagation

A multipath propagation medium contains several different paths by which energy travels from the transmitter to the receiver. If we consider firstly the case of a stationary receiver then we can imagine a "static multipath" situation in which several versions of the signal arrive sequentially at the receiver. The effect of the differential time delays is to introduce relative phase shifts between the component waves and superposition of the different components leads to either constructive or destructive addition (at one instant of time) depending upon the relative phases. Fig. 2.3 illustrates the two extreme possibilities. The resultant signal arising from paths A and B will be large because of constructive addition whereas that from paths A and C will be very small.

When either the transmitter or the receiver is in motion, we have a "dynamic multipath" situation in which there is a continuous change in the electrical length of every propagation path and thus the relative phase shifts between them change as a function of spatial location. Fig. 2.4 shows how the received amplitude (envelope) of the signal varies in the simple case when there are two incoming paths with a relative phase that varies with location. At some positions there is constructive addition whilst at others there is almost complete cancellation. In practice of course there are several different paths which combine in different ways depending on location and this leads to the more complicated signal envelope function shown in Fig. 2.2. The space-selective fading which exists as a result of multipath propagation is experienced as time-selective fading by a mobile receiver which travels through the field.

The time variations, or dynamic changes in the propagation path lengths, can be related directly to the motion of the receiver and indirectly to the Doppler effects that arise. In a practical case, the several incoming paths will be such that their individual phases, as experienced by a moving receiver will change continuously and randomly. The resultant signal envelope and RF phase will therefore also be random variables and it remains to devise a mathematical model to describe the relevant statistics. Such a model must be mathematically tractable and lead to results which are in accordance with the observed signal properties. For convenience we will only consider the case of moving receiver.

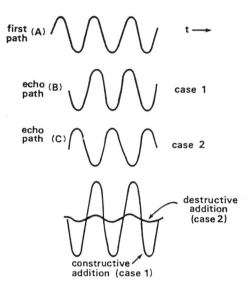

Fig. 2.3 Constructure and destructive addition of two transmission paths

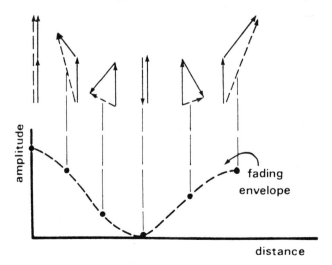

Fig. 2.4 Illustrating how the envelope fades as two incoming signals combine with different phases

2.3   Short-term fading

Several multipath models have been suggested to explain the observed statistical characteristics of the electromagnetic fields and the associated signal envelope and phase [1-3]. A model based on scattering is appropriate in general, and the most widely quoted and accepted is that due to Clarke [2]. It assumes that the field incident on the mobile antenna is composed of a number of plane waves of random phase, these plane waves being vertically polarised with horizontal angles of arrival and phase angles which are random and statistically independent. Furthermore the phase angles are assumed to have a uniform probability density function (PDF) in the interval (0,2π). This is reasonable at VHF and above, where the wavelength is short enough to ensure that small changes in path length result in significant changes in the RF phase. The PDF, p(α), for the horizontal angle of arrival of the plane waves is not specified by Clarke as it depends on the scattering characteristics of the environment and is likely to change over spatial distances greater than a few tens of wavelengths. A model such as this, based on scattered waves, allows the establishment of several important relationships describing the received signal, e.g. the first and second order statistics of the signal envelope and the nature of the frequency spectrum. Clarke's model, which is confined to the superposition of waves which travel only in the horizontal plane, has been generalised by Aulin [3] to allow the inclusion of vertically-polarised waves which do not necessarily travel horizontally. The results produced by Aulin differ from those of Clarke only in terms of the correlation properties (e.g. spectra) but not for the envelope and phase statistics. The differences are small and the Aulin model will not be considered here.

2.4   The scattering model

At every receiving point we assume the existence of N plane waves of similar amplitude. This is a realistic assumption in heavily built-up areas since the scattered components are likely to experience similar attenuation and there will be no dominant component. In certain situations a direct line of sight path may contribute a steady non-random component. This will alter the nature of the fading envelope and its statistics, but in this discussion we restrict analysis to the case of N equal-amplitude waves. The path-angle geometry for the nth scattered plane wave is shown in Fig. 2.5. If the transmitted signal is vertically polarised i.e. the electric field vector is aligned along the z-axis, the field components at the mobile are the electric field E_z, the magnetic field component H_x and the magnetic field component H_y. Generally, we deal only with the electric field component.

The assumption that the mobile received signal is of the scattered type with each component wave being independent, randomly phased and having a random angle of arrival, leads to the conclusion that the probability density function of the envelope is

$$p_r(r) = \frac{r}{\sigma^2} \ \exp\left(-\, r^2/2\sigma^2\right) \qquad \qquad ...(2.1)$$

which is a Rayleigh distribution. The corresponding cumulative distribution function P_r is

$$\text{prob}[r < R] = P_r(R) = 1 - \exp(- R^2/2\sigma^2) \qquad ...(2.2)$$

where σ^2 is the mean power.

2.5 Angle of arrival and signal spectra

If either the transmitter or receiver is in motion, the components of the received signal each experience a Doppler shift, the frequency shift being related to the spatial angle between the direction of arrival of that component and the direction of vehicle motion. For a vehicle moving at a constant speed v along the x-axis in Fig. 2.5 the Doppler shift f_n of the nth plane-wave component is

$$f_n = \frac{v}{\lambda} \cos \alpha_n \qquad 1 \le n \le N \qquad ...(2.3)$$

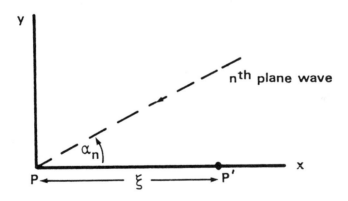

Fig. 2.5   Path-angle geometry for the nth plane wave

It can be seen that component waves arriving from ahead of the vehicle experience a positive Doppler shift (maximum value $f_m = v/\lambda$) whilst those arriving from behind the vehicle have a negative shift.

If n is large, the fraction of the incident power contained within a spatial angle between α and $\alpha + d\alpha$, for an omnidirectional antenna is $p(\alpha)d\alpha$; if the antenna has a gain $G(\alpha)$ the corresponding value is $G(\alpha)p(\alpha)d\alpha$. Equating this to the incremental power determined through the relationship between Doppler shift and spatial angle, viz

$$f(\alpha) = f_m \cos\alpha + f_c \qquad\qquad ...(2.4)$$

we obtain the frequency spectrum S(f) of the received signal as

$$S(f) \mid df \mid = \{G(\alpha)p(\alpha) + G(-\alpha)p(-\alpha)\} \mid d\alpha \mid \qquad ...(2.5)$$

We can obtain an expression for S(f) in the case of a vertical monopole by assuming p(a) to be uniformly distributed in the angular range $(-\pi, \pi)$. This simple assumption leads to an RF spectrum given by:

$$S_{E_z}(f) = \frac{1.5}{\pi f_m}\left[1 - \frac{(f - f_c)^2}{f_m}\right]^{0.5} \qquad ...(2.6)$$

This, together with the corresponding baseband spectrum is shown in Fig. 2.6(a).

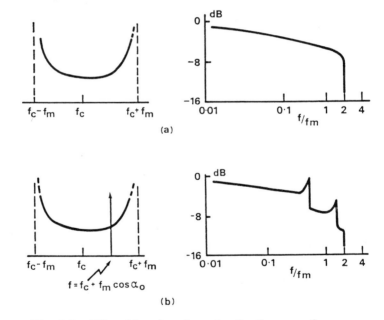

(a)

(b)

Fig. 2.6   RF and baseband spectra for the case of
(a) isotropically scattered signals, (b) when there is a dominant component

If a dominant component exists in the incoming signal then this has a substantial influence on the signal spectrum. Such a component arriving at an angle α_0 gives rise to a spectral line at $f_c + f_m \cos \alpha_0$ in the RF spectrum and two additional peaks at $f_m (1 + \cos \alpha_0)$ in the baseband spectrum as shown in Fig. 2.6(b).

In the time domain, as we have already seen, the effects of the randomly phased and Doppler-shifted multipath signals appear in the form of a fading envelope.

2.6 Fading envelope statistics

The fading envelope directly affects the performance of radio receivers. The Rayleigh fading envelope only occasionally experiences very deep fades, for example 30 dB fades occur for only 0.1% of the time. To specify some of the constraints in a quantitative manner we can derive some parameters of the fading envelope that have direct relevance to system performance. In particular we are interested in how often the envelope crosses a specified signal level (the level-crossing rate), and the average duration of a fade below that specified level.

2.6.1 Level-crossing rate

The level crossing L rate at a specified signal level R is defined as the average number of times per second that the signal envelope crosses the level in a positive-going direction. Mathematically this can be expressed as [4],

$$L(R) = \int_0^\infty \dot{r}\, p(R, \dot{r})\, d\dot{r} \qquad \qquad ...(2.7)$$

where $p(R,\dot{r})$ is the joint p.d.f. of R and r, and a dot indicates the time derivative.

For a vertical monopole the level crossing rate becomes

$$L(R) = \sqrt{2\pi}\, \rho \exp(-\rho^2) \qquad \qquad ...(2.8)$$

where $\rho = R/R_{rms}$, the ratio between the specified level and the r.m.s. amplitude of the fading envelope.

The normalised level crossing rate for a vertical monopole is plotted in Fig. 2.7. The maximum rate occurs at $\rho = -3$ dB and decreases as the level is lowered or raised. We recall that since the level crossing rate is specified for a positive direction then when a low level is set the signal remains above this level for most of the time; a similar argument applies when a level much greater than the RMS value is selected. At $f_c = 900$ MHz and a vehicle speed of 48 kph, $f_m = 40$ Hz, thus the level crossing rate L_R is 39 per second at $\rho = -3$ dB.

2.6.2 Average fade duration

The average duration of fades below the specified level R can be found from the relationship

$$\text{Average fade duration} = \frac{\text{prob}[r < R]}{L(R)} \qquad ...(2.9)$$

From eqn (2.2), prob[r < R] can be evaluated as $1 - \exp(-\rho^2)$, so substituting for L(R) from eqn (2.8) gives the average fade duration for a vertical monopole as

$$\tau(R) = \frac{\exp(\rho^2) - 1}{\rho f_m \sqrt{2\pi}} \qquad ...(2.10)$$

and this is plotted in Fig. 2.8.

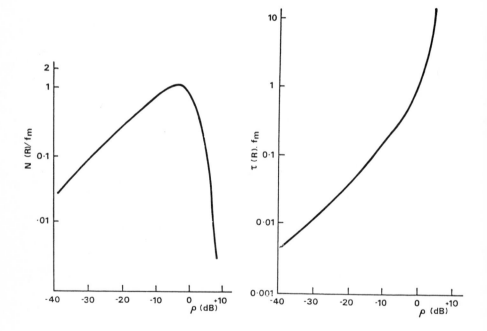

Fig. 2.7   Normalised level-crossing rate for a vertical monopole under conditions of isotropic scattering

Fig. 2.8   Normalised average duration of fades under conditions of isotropic scattering

2.7 Spatial correlation of field components

In mobile radio systems, especially at VHF and above, the effects of fading can be combatted by the use of space diversity techniques either at the base station or the mobile provided that antennas can be spatially separated far enough to ensure that the signal envelopes exhibit a low correlation. The principles of diversity will not be considered here; we will merely indicate the way in which decorrelated signals can be obtained. To obtain the spatial correlation function for the electric field we consider the field incident at a point P' a distance ξ along the axis from the point P as shown in Fig. 2.5. The phase of the nth component wave at P' will be $(\emptyset n + k\xi \cos\alpha_n)$ where k $(= 2\pi/\lambda)$ is the phase shift factor in free space. The crosscorrelation between the complex signals at the two points P and P' can be obtained via the autocovariance function of the electric field. If we assume statistical independence of $\emptyset$ and α and also consider the case of isotropic scattering i.e. assume that $p(\alpha) = 1/2\pi$, then it follows that

$$R_{E_z} = NE_o \int_{-\pi}^{+\pi} p(\alpha) \exp (jk\xi\cos\alpha)\, d\alpha$$

$$= NE_o^2 J_o (k\xi)$$

where $J_o(.)$ is the zero-order Bessel function of the first kind.

It can then be shown that to a good approximation the normalised autocovariance function of the envelope is equal to the square of the normalised autocovariance function of the complex random field. Thus

$$\rho \mid E_z \mid = J_o^2 (k\xi) \qquad \qquad ...(2.11)$$

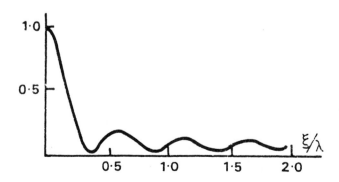

Fig. 2.9   Autocovariance function for the envelope
of the electric field

This function is plotted in Fig. 2.9 for the case of an isotropically scattered field. There is rapid decorrelation, showing that space diversity can be implemented at the mobile end of the link where the assumption of isotropic scattering is approximately true.

On the other hand, it is clear that at base station sites the assumption of isotropic scattering is unlikely to hold. Base station sites are deliberately chosen to be well above local obstructions to give the best coverage of the intended service area and the scattering objects which produce the multipath effects are located principally in a small area surrounding the mobile. The reciprocity theorem applies, of course, in any linear medium but this should not be taken to imply that the spatial correlation distance at one end of the radio path is the same as it is at the other.

It seems intuitively obvious that antennas have to be further apart at base station sites than at mobiles to obtain the same correlation and indeed this is the case. It is also apparent that whereas at mobiles the assumption of isotropic scattering (from scatterers which surround the mobile uniformly) leads directly to the conclusion that the correlation between the electric field at two receiving points is a function of their separation only, this is not the case at base station sites. Here, scattering is not isotropic and the correlation between the electric field at two receiving points is a function of both their separation and the angle between the line joining them and the direction to the mobile.

A further possibility for obtaining decorrelated signals is to separate the receiving points vertically rather than horizontally. Clarke's model, which is two-dimensional, cannot cope with this situation but Aulin's model can do so. There have been experimental investigations of this possibility [5,6] which have demonstrated its feasibility at base station sites. In practice at 900MHz the required separation is about 12λ when the mobile to base distance is about 1.5km [6].

2.8 Random FM

The discussion so far has concentrated on the properties of the signal envelope, but it was mentioned earlier that the apparent frequency of the signal varies with time. The phase and its time derivative vary in a random manner and the signal therefore exhibits random frequency variations known as random frequency modulation. The characteristics of this random FM can be described in terms of its probability density function and power spectrum and appropriate expressions have been obtained in reference 4.

For the electric field component the probability density function of the random FM can be written as

$$\dot{p}(\theta) = \frac{1}{\omega_m \sqrt{2}} \left[1 + 2 \left| \frac{\theta}{\omega_m} \right|^2 \right]^{-3/2} \qquad \text{...(2.12)}$$

In contrast to the well-defined band-limited power spectrum of the envelope (the Doppler spectrum) there is a finite probability of finding the frequency of the random FM at any value, although the larger frequency excursions are caused by the deep fades and therefore occur only rarely. For a 20 dB fade the significant frequency deviations occupy a bandwidth of about $10f_m$, this decreases to f_m when the signal level is at its RMS value.

The power spectrum of the random FM may be derived as the Fourier transform of the autocorrelation function of θ. The mathematics is rather involved but leads to the one-sided power spectrum shown in Fig. 2.10. Above an angular frequency of $2\omega_m$ the spectrum falls off as $1/f$. Thus $2\omega_m$ can be regarded as a "cut-off" frequency, both for the random FM and for the signal spectrum in Fig. 2.6.

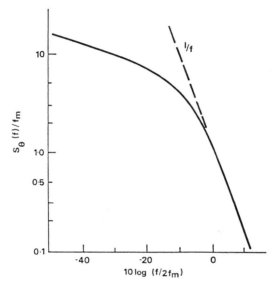

Fig. 2.10 Power spectrum of random FM plotted as relative power on a normalised frequency scale

2.9 Frequency-selective fading

The earlier discussion has been concerned with describing the envelope and phase variations of the signal received at a moving vehicle when an unmodulated carrier is radiated by the base station transmitter. The question now arises as to how adequate is this description of the channel when real signals, which occupy a finite bandwidth, are radiated. It is clear that in practice we need to consider the effects of multipath propagation on these signals and as an example we consider the case of two frequency components within the message bandwidth. If these frequencies are close together then the different propagation paths within the multipath medium have approximately the same electrical length for both components and their amplitude and phase variations will be very similar. This is the "narrowband" case. As the frequency separation increases, however, the behaviour at one frequency tends to become uncorrelated with that at the other frequency because the differential phase shifts along the various paths are quite different at the two frequencies. The extent of the decorrelation depends on the spread of time delays since the phase shifts arise from the excess path lengths. For large delay-spreads the phases of the incoming components can vary over several radians even if the frequency separation is quite small. Signals which occupy a bandwidth greater than that over which spectral components are affected in a similar way will become distorted since the amplitudes and phases of the various spectral components in the received version of the signal are not the same as they were in the transmitted version. The phenomenon is known as frequency-selective fading and the bandwidth over which the spectral components are affected in a similar way is known as the coherence, or correlation bandwidth.

The fact that the lengths of the individual propagation paths vary with time due to motion of the vehicle gives us a method of gaining further insight into the propagation mechanism, since the changing time of arrival suggests the possibility of associating each delayed version of the transmitted signal with a physical propagation path. Indeed, if a number of distinct physical scatterers are involved, it may be possible to associate each scatterer with an individual propagation path. It is not possible, however, to distinguish between different paths merely by considering the difference between the times of arrival, the angular direction of arrival also has to be taken into account. If we consider only single-scattered paths then all scatterers with the same path length can be located on an ellipse with the transmitter and receiver at its foci. Each time delay between transmitter and receiver defines a confocal ellipse as shown in Fig. 2.11. If we consider scatterers located at A, B and C, then we can distinguish between paths TAR and TBR, which have the same angle of arrival, by their different time delays and between TAR and TCR which have the same time delay, by their different angles of arrival.

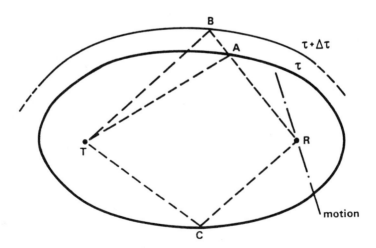

Fig. 2.11 Path geometry for single scattering

The angles of arrival can be determined by means of Doppler shift. As we have already seen, whenever the receiver or transmitter is in motion the received RF signal experiences a Doppler shift, the frequency shift being related to the cosine of the spatial angle between the direction of arrival of the wave and the direction of motion of the vehicle. If, therefore, we transmitted a short RF pulse and measured both its time of arrival and Doppler shift at the receiver we could identify the length of the propagation path and the angle of arrival. Of course, there is a left/right ambiguity inherent in the Doppler shift measurement but this could be resolved, if necessary, by the use of directional antennas.

An important and instructive feature of Fig. 2.11 is that for a particular receiver location a suitably scaled diagram with several confocal ellipses can be produced in the form of a map overlay. Co-ordinated use of this overlay, together with experimental results for the location in question allows the identification of significant single scatterers or scattering areas, and gives an indication of the extent of the contribution from multiple scattering.

2.10 Channel characterisation

Signals that have suffered multipath propagation constitute a set of randomly attenuated, delayed, and phased replicas of the transmitted RF signal. The resultant, sensed at the receiver, is therefore the superposition of contributions from the various individual paths and on this basis it is reasonable to describe the radio channel in terms of a two-port filter with randomly time-varying transmission characteristics. Since the signals at the input and output of the equivalent filter can be represented in either the frequency or time domains there are four possible input-output functions which can be used to describe the behaviour of the radio channel [7].

Consider the time-domain description of a deterministically time-variant channel, expressed in terms of the impulse response of the equivalent filter. For a time-varying channel the impulse response is also time variant. In the notation of complex envelopes of real bandpass waveforms the output $z(t)$ is related to the input $y(t)$ and the impulse response $h(t,\tau)$ by the expression

$$z(t) = \int_{-\infty}^{\infty} y(t-\tau) \ h(t,\tau) \ d\tau \qquad \qquad ...(2.13)$$

A circuit model of the complex convolution represented by eqn. (2.13) is shown in Fig. 2.12 in the form of a tapped delay-line transversal filter. From a physical viewpoint the multipath is seen to arise from a continuum of stationary elemental scatterers distinguished in path lengths by their elemental delays $d\tau$. Each scatterer produces gain fluctuations $h(t,\tau) \ d\tau$ which are time dependent. Modelling the time-variant impulse response of the channel in the form of a densely-tapped delay line provides added insight into the physical mechanisms causing multipath, since it explicitly shows that the received signal comprises delayed, attenuated replicas of the input signal and it demonstrates the nature of multipath arising from scatterers with different path lengths.

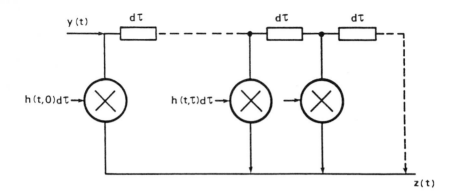

Fig. 2.12 Tapped delay line model of the channel

In a similar way we can provide a direct illustration of the Doppler shift phenomenon by characterisation in terms of $H(f,\upsilon)$, a dual function of $h(t,\tau)$ in the frequency/frequency-shift domain. This is defined in terms of the complex spectra at input and output by the superposition in Doppler shifts, υ.

$$Z(f) = \int_{-\infty}^{\infty} Y(f-\upsilon) \; H(f-\upsilon,\upsilon) \; d\upsilon \qquad \qquad ...(2.14)$$

Characterising a channel using $H(f,\upsilon)$ necessitates measurement over a suitably wide frequency range in order to study the frequency selective effects caused by paths of varying lengths.

Another characterising function $T(f,t)$ describes the response of the channel to cissoidal excitations in the frequency band of interest. In effect this is the time-variant transfer function of the channel, defined by

$$z(t) = \int_{-\infty}^{\infty} Y(f) \; T(f,t) \; \exp(j2\pi ft) \; df \qquad \qquad ...(2.15)$$

The effects of multipath propagation are now observed as time-selective fading of $T(f_1,t)$ for transmission of a single frequency f_1. $T(f,t)$ is the function we used earlier to characterise narrowband radio channels, the underlying assumption being that the fading character at a single frequency is representative of the actual signals received. However, $T(f,t)$ does not give direct physical insight into the multipath phenomenon since it only displays the manifestation of multipath as a fading envelope of the received signal. It can therefore be seen that while h(t,τ) directly distinguishes paths of different lengths, H(f,υ) can directly identify different Doppler shifts, each shift being associated with the angle of arrival of a scattered component path.

A system function which can simultaneously provide an explicit description of the multipath in both the time-delay and Doppler-shift domains can be introduced on the following basis. The Doppler spectra are embedded in the time-varying complex envelope of h(t,τ) and a system function S(τ,υ) can therefore be envisaged such that while it retains explicitly the time delays of h(t,τ) the Doppler spectra emerge more directly from the complex envelope by defining S(τ,υ) as the spectrum of h(t,τ) i.e.

$$h(t,\tau) = \int_{-\infty}^{\infty} S(\tau,\upsilon) \; \exp(j2\pi\upsilon t) \; d\upsilon \qquad \qquad ...(2.16)$$

The input-output relation then becomes

$$z(t) = \int_{-\infty}^{\infty} \int_{-\infty}^{\infty} y(t-\tau) \; S(\tau,\upsilon) \; \exp(j2\pi\upsilon t) \; d\upsilon \; d\tau \qquad \qquad ...(2.17)$$

and represents the sum of delayed, weighted and Doppler-shifted signals. Thus signals arriving during the infinitesimal delay range τ to τ + dτ and having Doppler shifts in the range υ to υ + dυ have a differential gain of S(τ,υ) dτ dυ.

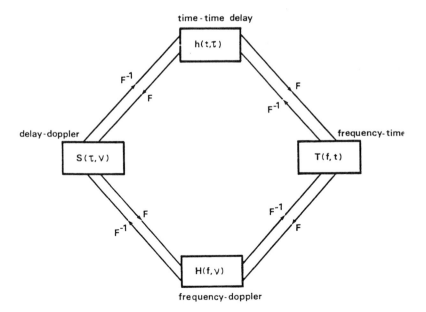

Fig. 2.13 Relationships between system functions in different domains

Significant symmetric relationships emerge on inspection of the four system functions [8] and are shown in Fig. 2.13. The Fourier transform and inverse Fourier transform relationships between these functions are indicated by arrowheads with F and F^{-1} respectively. Thus measurements in one domain can be related to parameters derived from a system function in another domain.

In practice, multipath radio channels have random, rather than deterministic, characteristics and it is then necessary to consider the system functions as stochastic processes which can be specified by their statistical moments. In all practical cases this is confined to a study of the mean and correlation functions of the two-dimensional system functions; for channels with Gaussian statistics the mean and correlation functions provide an exact description. An extension of the system functions to include the random behaviour has a twofold advantage. Firstly, we retain the description in terms of a two-port filter, thereby allowing the autocorrelation function of the channel output to be obtained from a knowledge of the autocorrelation of the input-output system function. Secondly, the behaviour of the correlation functions aids understanding of the physical scattering of radio waves. Statistical classification of the channel on the basis of the time and frequency statistics, i.e., correlation functions, provides vital clues about the nature of the scattering sources. This aspect of channel characterisation is covered in references 7 to 11.

2.11   Channel sounding techniques

The symmetrical relationships between the various input-output channel functions via the Fourier and inverse-Fourier transforms imply that a characterisation in either the time or frequency domain is adequate. For narrowband channels, sounding by means of an unmodulated CW carrier leads to characterisation via the T(f,t) function. However for wideband (frequency-selective) channels other methods have to be used. One possibility is to simultaneously (or sequentially) transmit several spaced carriers and to receive each one via a narrow-band receiver at the mobile. This leads directly and conveniently to a measurement of the coherence bandwidth via the fading characteristics of the different frequencies, but measurements of time-delay and delay-spread are less easily extracted from the results. This technique has been used [12] and although there are some advantages, in general it is not as flexible as the alternative technique in the frequency/time domain of using a swept CW carrier i.e. a chirp signal.

Channel sounding techniques in the time/time-delay domain using a wideband RF pulse, or its equivalent, amount to measuring the channel impulse response and explicitly show the time-delay multipath characteristics. Coherent detection of the in-phase and quadrature components of the RF pulse also allow the Doppler shifts associated with each time delay to be extracted after suitable signal processing, and hence the angles of arrival can be determined. Measurements in the frequency-time domain using a swept CW carrier (i.e. a "chirp" signal) provide essentially the same information, but not explicitly in terms of the time-delay or Doppler-shift multipath characteristics.

2.12 Small-area characterisation

It is clear that the characteristics of radio propagation in urban areas depend largely on factors such as the nature of the terrain, the height and density of buildings and the street orientation; these characteristics can be studied by observing the features of measured scattering functions. For example, Fig. 2.14 shows a scattering function measured in a suburban area of city, the particular road in this case being almost radial with respect to the transmitter. Two-storey detached houses are evenly spaced along this street on one side while the opposite side has fewer but larger dwellings. Echoes in the scattering function are identified below by the co-ordinates (x,y) where x is the excess time delay in μs and y is the Doppler shift in Hz. Thus the large amplitude echo at (0.1, -3.8) represents the shortest path (in the absence of an LOS path) in this mid-block section of the street, and this scattered component is expected in view of the near-radial street orientation with respect to the transmitter. The presence of strong echoes with near-maximum Doppler shifts leads to the conclusion that the majority of the scattered waves arrive from ahead of and behind the vehicle rather than from the 90° angular directions. Scattering from the flat sides of houses in the immediate vicinity of the mobile is indicated by the presence of echoes within the rectangle defined by the co-ordinates (0.1, 0) to (0.3, 2.5). The elevated houses lying ahead contribute scattered components with excess delays up to 0.5μs. Weaker echoes with longer delays and negative Doppler shifts are due to non-local reflections from large buildings some distance away. Clearly the scattering in this residential area is localised to the immediate environment and depends on the street orientation and sloping terrain.

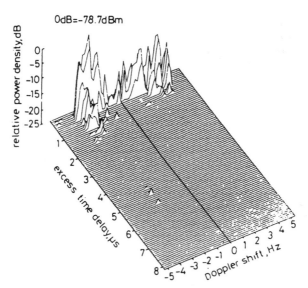

Fig. 2.14 Measured scattering function in a
radial suburban street

In certain environments, isolated large buildings can play a significant role in determining the multipath characteristics. Fig. 2.15 shows a scattering function for another road and this represents a more extreme case of suburban multipath. The direct LOS path is blocked by a large building, and other non-local features that influence propagation in this area include some high-rise residential blocks. A significant feature at this location is the wide spread of time-delays due to contributions from the large non-local scatterers. Due to the complex topography it is not possible to explain all echoes on a basis of single scattering. In general, the local scattering includes all angles of arrival for the radiowaves due to the circumferential orientation of the street. Furthermore, the scattering is uncorrelated for different time delays, this being evident from the dissimilar Doppler spectra at different delays. The presence of large scatterers in suburban areas gives multipath radio propagation an inhomogeneous character and large reflected components, when shadowed due to motion of the vehicle relative to the local topography, can drastically alter the resultant scattering.

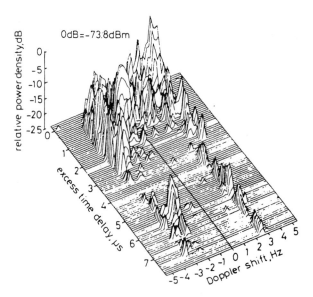

Fig. 2.15   Measured scattering function in an area
having extreme multipath conditions

In densely built up city areas propagation is mainly by means of scattering. The environment is more uniform than in suburban areas, although echoes from large buildings can sometimes be identified. Nevertheless, significant differences can be observed between data collected at street intersections and that collected in mid-block areas. Street orientation influences the scattering pattern in mid-block areas, but at intersections, radio waves scattered by buildings along both streets arrive unobstructed at the mobile receiver and street orientation with respect to the transmitter does not influence the angular distribution of scattered paths. In certain situations the non-local scattering can also change.

2.13 Statistical characterisation

The discussion in the previous section has illustrated the role of the scattering function in the study of multipath radio propagation. Assessment of communication systems can be aided by viewing this, and other system functions, as statistical input-output descriptors of the two-port channel model. Because the power-delay profiles represent a statistical function it is possible to extract parameters that indicate the severity of the multipath. These parameters are the first and second central moments of the average power delay profile, the average delay and the delay spread, respectively.

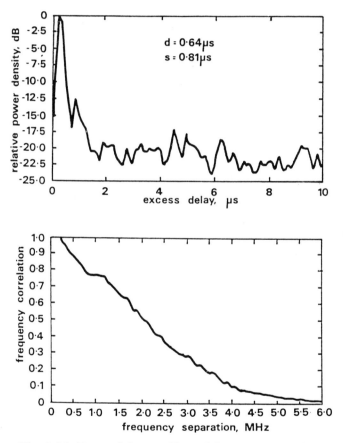

Fig. 2.16 Power delay profile and frequency correlation function in an urban area

In suburban areas, multipath is predominantly due to local scattering from residential houses and the time delay dispersion is usually less than 1µs. The power delay profile and frequency correlation function for a typical suburban street are shown in Fig. 2.16, the average delay being 0.64μs and the delay spread, 0.81μs. The delay dispersion is small and this leads to a smoothly varying frequency correlation function and a reasonably large coherence bandwidth of 2 MHz. To obtain some indication of the likely overall performance of communication systems in suburban areas it is necessary to consider a more extreme multipath situation such as that represented by the power delay profile and frequency correlation function in Fig. 2.17 which correspond to the scattering function of Fig. 2.15. The average delay and the delay spread are more than double the 'typical' values and the frequency correlation function, although it decreases rapidly at first, contains a number of oscillatory variations. The periodicity in the fine oscillatory structure in the correlation function is determined by the echo at an excess delay of 5.5μs in the power delay profile, and a coherence bandwidth of only 200 kHz indicates that the radio channel is highly frequency-selective. The difference between typical and extreme multipath conditions highlights the variability of propagation conditions in suburban areas.

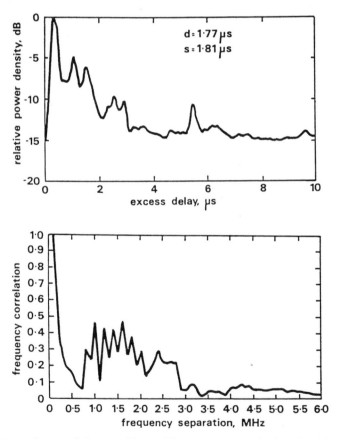

Fig. 2.17   Power delay profile and frequency correlation function
in a suburban area under extreme multipath conditions

The multipath characteristics in the dense high rise localities are representative of typical urban areas. In these localities the scattering is generally stable due to the homogeneous environment, and consequently the multipath parameters are less variable. A power-delay profile taken close to an intersection is shown in Fig. 2.18. The echoes due to local scattering decrease in strength except for the large echo at an excess delay of 2ms. The presence of such an echo appears in the form of oscillations in the frequency correlation function. Using a definition of coherence bandwidth in which the frequency separation for 0.5 correlation is taken, it is clear that three different values are observed. Obviously the smallest frequency separation indicates the actual coherence bandwidth but nevertheless it demonstrates that measurement of coherence bandwidth using spaced tones is not a satisfactory method unless measurements are carefully repeated for different frequency separations. In the literature [4,13] a smoothly decreasing echo power distribution is often assumed implying a smooth decorrelation due to the inherent relationship between these functions [8]. Clearly these assumptions are unrealistic for multipath propagation in certain suburban, and most urban

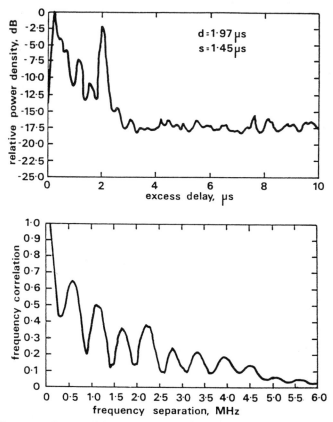

Fig. 2.18 Power delay profile and frequency correlation function in an urban area

areas. More extreme multipath characteristics in urban areas can be found [10], particularly in industrial areas. Echoes with large excess delays are often a salient feature of the power delay profile, these echoes being due to non-local scattering from large buildings some distance away. It is interesting to note the resemblance of these multipath characteristics to those in suburban locations influenced by remotely sited high rise buildings, because in both areas this environment represents an extreme case. The first minimum in the frequency correlation function often indicates significant decorrelation, although not enough to reach the value of 0.5. However it could be argued that this is indicative of the actual coherence bandwidth and that the value obtained by considering decorrelation to 0.5 is deceptively optimistic. The very substantial influence of strong delayed components is obvious in these cases since components at a certain excess delay produce corresponding oscillatory variations in the frequency correlation function.

2.14 Large-area characterisation

In the previous sections we have discussed parameters suitable for channel characterisation. It is evident that the multipath characteristics in urban areas are grossly non-stationary in nature, so that experimental data collected at a given location provides parameters that, strictly, describe propagation only at that location, i.e. provide a small-area characterisation. Extension from local to global characterisation is therefore not a trivial matter. Nevertheless it is very important, because designers of radio communication systems intended to operate in the presence of multipath wish to know, given a certain performance criterion, at what percentage of locations within the service area the equipment will be satisfactory and what can be done to improve matters.

We therefore consider briefly the question of large-area characterisation and attempt to provide a bridge between the local and the global situation. It has been observed that many radio channels have fading characteristics that are stationary over short periods of time (or over short distances of travel). These channels are not stationary in the strict sense but can be regarded as stationary in the wide sense i.e. weakly stationary. Wide-sense stationary (WSS) channels have the property that the channel correlation function are invariant under a translation in time.

Over spatial distances of the order of a few tens of wavelengths the descriptive statistics of the mobile radio channel are wide-sense stationary. As the mobile receiver moves from one locality to another, changes in the dominant features of the environment are manifested in the form of nonstationarity in the statistical characteristics of the multipath and it is reasonable to study the non-stationary behaviour by observing the variability of these parameters over a large, but characteristically similar, area.

The approach therefore is to obtain WSS channel descriptors in the form of average power/time-delay profiles for a large number of relatively small areas and to observe the variation in the parameters of these descriptors. In general, the average echo-power versus time-delay represents a statistical distribution of echo power over the small area. Therefore, the average excess delay and delay-spread, are relevant statistical parameters of the channel. These parameters are insensitive to the shape of the power-delay profile for homogeneous environments due to the predominantly scattered mode of radio propagation. However this is not the case in the presence of strong reflected components that have large excess delays, this situation being evidenced in practice by spikes in the power-delay profile at the longer values of time delay.

Histograms of the average delay and delay spread in suburban areas are shown in Fig. 2.19. For a significant percentage of locations the values of these two parameters greatly exceed the values typical of the areas under consideration. For example, at 22% of locations the average delay is more than $\mu_d + \sigma_d$ (1.4μs). The extreme values, which correspond to the 'tails' of the statistical distribution of the parameters, occur predominantly in areas with high rise buildings. These 'tails' exhibit the same behaviour for the two parameters since they consistently constitute more than 20 percent of the locations. The bimodal nature of the histograms, emphasizing low and high values of delay and delay-spread reflects the grossly inhomogeneous composition of the suburban environment. System designs based on typical or mean values of the parameters are

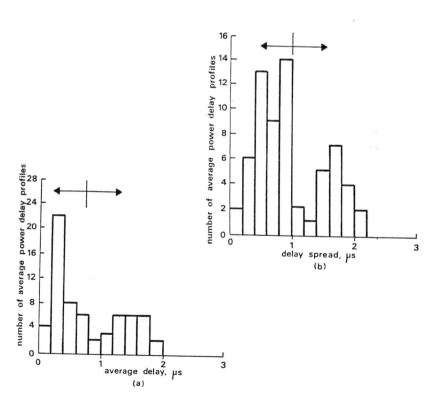

Fig. 2.19 Histograms of average delay and delay
spread, typical of suburban areas

therefore unlikely to provide the desired overall performance due to the large variability of the channel parameters over the service area. In general the more typical residential areas display less time-delay dispersion than suburban areas influenced by non-local high-rise buildings.

A histogram of the coherence bandwidth of the channel (as defined by a 0.5 correlation coefficient) is shown in Fig. 2.20. The lower extreme signifies the worst multipath dispersion which, of course, manifests itself in the form of frequency selectivity. For multipath resulting mainly from scattering, the coherence bandwidth B_c is inversely proportional to the delay spread s.

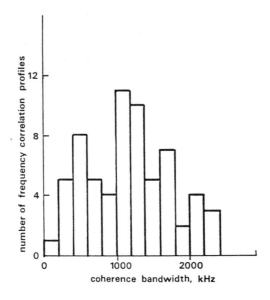

Fig. 2.20   Histogram of coherence bandwidth
typical of suburban areas

Histograms of average excess delay and delay spread for urban areas are shown in Fig. 2.21. As far as average delay is concerned the very low values are found to a greater extent in dense high-rise areas. Moderate values for the excess delay have a tendency to occur in all types of urban environment.

The low to moderate values of delay spread in Fig. 2.21 have been mainly derived from observations in dense high rise surroundings while the much higher values occur only in commercial and industrial urban areas. The variance of the delay spread is smaller than in suburban areas due to the greater homogeneity of the urban environment and the extreme value, $\mu_s + \sigma_s = 2\mu s$, is exceeded in only 7 percent of the locations. This is comparable to the fluctuations induced by the statistical sample size. However, the average excess delays exceed $\mu_d + \sigma_d = 2\mu s$ in 20 percent of the locations and therefore require consideration in the assessment of automatic vehicle location systems.

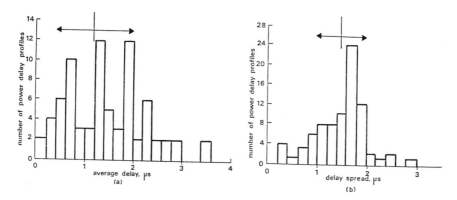

Fig. 2.21 Histograms of average delay and delay
spread typical of urban areas

2.15 Conclusions

A test transmission in the form of an unmodulated RF carrier provides an adequate channel characterisation for the case when the intended information transmission occupies a narrow bandwidth and frequency-selective fading is not likely to be a problem. In this case, the envelope is Rayleigh distributed when measured over distances of a few tens of wavelengths and the phase is uniformly distributed over $(0, 2\pi)$. A two-dimensional isotropic scattering model enables all the principal properties of the received signal to be predicted with a high degree of accuracy.

A narrow RF pulse, or its equivalent, is a favoured test signal when a wideband characterisation is appropriate. In this case small scale channel descriptors in the form of the average power-delay profile provide useful information about mobile radio channels and the variability of the statistical parameters, each extracted from a WSS channel representation, is an appropriate large-scale descriptor in a two-stage characterisation of the transmission medium. Extreme values of average delay and delay spread occur quite frequently in suburban areas; they are most common in inhomogeneous environments with high-rise residential buildings, irrespective of street orientations. Urban areas show comparable extreme values though less frequently in the case of the delay spread. The outer fringe of the urban environment often represents the worst multipath, being influenced to a greater extent by remote but tall buildings some distance away. The effects of inhomogeneous scattering are also evident in the statistical distributions for the coherence bandwidths. Variability of the parameters is greater in suburban areas although the extreme values are comparable. This large scale characterisation can be applied to an evaluation of radio systems [15] and the performance bounds then emerge in terms of the percentage of locations where the given performance measure is not exceeded.

2.16   References

1.   OSSANA, J.F., Jr.: 'A model for mobile radio fading due to building reflections: theoretical and experimental fading waveform power spectra', Bell. Syst. Tech. J., 1964, Vol. 43, pp.2935-2971

2.   CLARKE, R.H.: 'A statistical theory of mobile radio reception', Bell. Syst. Tech. J., 1968, Vol.47, pp.957-1000

3.   AULIN, T.: 'A modified model for the fading signal of a mobile radio channel', IEEE Trans., 1979, VT-28, pp.182-203

4.   JAKES, W.C., Jr. (Ed): 'Microwave mobile communications' (John Wiley Interscience, New York, 1974)

5.   RHEE, S.B., and ZYSMAN, G.I.: 'Results of suburban base-station spatial diversity measurements in the UHF band', IEEE Trans., 1974, COM-22, pp.1630-1634

6.   ADACHI, F., FEENEY, M.T., WILLIAMSON, A.G., and PARSONS, J.D.: 'Correlation between the envelopes of 900MHz signals received at a mobile radiobase-station site', Proc. IEE, 1986, Vol. 133, Part F, pp.506-512

7.   BELLO, P.A.: 'Characterisation of randomly time-variant linear channels', IEEE Trans., 1963, Vol.CS-11, pp.360-393

8.   PARSONS, J.D., and BAJWA A.S.: 'Wideband characterisation of fading mobile radio channels', Proc. IEE, 1982, Vol.129, Part F, pp.95-101

9.   COX, D.C.: '910MHz urban mobile radio propagation: multipath characteristics in New York City', IEEE Trans., 1973, COM-21, pp.1188-1194

10.   BAJWA, A.S.: 'Wideband characterisation of UHF mobile radio propagation in urban and suburban areas', Ph.D. thesis, Department of Electronic and Electrical Engineering, University of Birmingham, 1979

11.   KENNEDY, R.S.: 'Fading dispersive communication channels' (John Wiley Interscience, New York, 1969)

12.   MATTHEWS, P.A., MOLKDAR, D., and RASHIDZADEH, B.: 'Measurement, description and modelling of the UHF terrestrial mobile radio channel', Proc. 3rd Int. Conf. on Land Mobile Radio, Cambridge, Dec. 1985 (IERE Conf. Publ. No.65 pp.119-125)

13. BELLO, P.A., and NELIN, B.D.: 'The effect of frequency fading on the binary error probabilities of incoherent and differentially coherent matched filter receivers', IEEE Trans., 1963, CS-11, pp. 170-186

14. COX, D.C., and LECK, R.P.: 'Distribution of multipath delay spread and average excess delay for 910 MHz urban mobile radio paths', IEEE Trans., 1975, AP-23, No.2, pp. 206-213

15. BAJWA, A.S., and PARSONS, J.D.: 'Large area characterisation of urban UHF multipath propagation and its relevance to the performance bounds of mobile radio systems', IEE Proc, 1985, Vol.132, Part F, pp. 99-106

Chapter 3

Modulation techniques: analog and digital

Raymond C V Macario

3.1 Introduction

Modulation is the process whereby the message information is added to the radio carrier. The carrier wave was depicted in the first diagram in the book, namely Fig. 1.1. This is the carrier which attempts to get to and from the mobile, subject to the propagation losses and multipath effects described earlier. Adding modulation widens the bandwidth of the signal. Ideally, a carrier has no bandwidth, but in practice, drift, noise and propagation produce observable bandwidth. Modulation produces sidebands, either on one side, or both sides of the nominal carrier frequency, usually having a distribution of amplitude over the assigned bandwidth. In effect one could say whatever the modulation, one therefore has a situation whereby one has a concentrated group of closely knit carriers or signals coming to and going from the mobile, *whether one specifies analog or digital modulation.*

The question one needs to ask, or solve, is what is the best way of arranging this new group of signals; analog FM or continuous phase MSK, for example? This chapter is not for the purpose of choosing or recommending the best grouping of this package of sinusoidal signals. To some extent engineering history has made its own decisions. For example as soon as J. R. Carson [1] drew attention to the fact that multiplying a carrier by the modulation (speech or tone) produces a pair of sidebands the world launched into a scene of rapid growth AM and SSB broadcasting and also concepts of reduced carrier.

In fact, it took the superheterodyne inventor E. H. Armstrong [2] to show that the sidebands, surrounding the carrier position, can be arranged quite differently if necessary and the important merits of FM, namely; noise reduction, sound quality, impulsive noise masking and no agc requirements could be achieved. FM has since taken over from AM, and narrow band FM modulation dominates the mobile market.

However, recently the world has been moving to digital operation and the realization that the control and switching between users can be performed much better by digital operation. Why not use digital modulation throughout the system [3]? Can the signalling group around the carrier be arranged as efficiently as in the case of FM? What one finds is the converging of all forms of modulation to form the transmitted signal package; the main difference between analog and digital modulation to some extent being the extensive use of digital signal processing prior to modulation and often after demodulation in the digital case.

3.2 The bandwidth problem

The bandwidth problem arises because, either one divides the users into a well defined frequency division multiple access (FDMA) arrangement, or one divides the users into time division multiple access (TDMA) time slots which requires they transmit their information smartly at well defined intervals in the total assigned bandwidth.

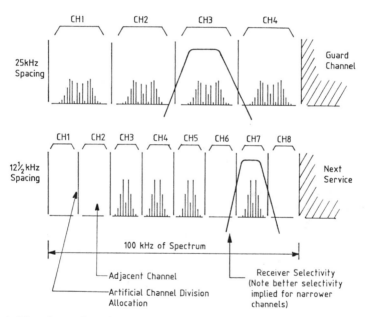

Fig. 3.1 The channel assignment arrangement in a single carrier per carrier system showing the sidebands within each channel, for two cases

In an FDMA system spectrum partitioning is arranged and, as shown in Fig. 3.1, our user's carrier assignment and cluster of modulation sidebands needs to be accurately placed in the assigned channel. Sidebands at the edge of the assigned channel are required to be at least 60 dB below the unmodulated carrier amplitude.

This situation applies whether the modulation system is analog or digital. For example, if the system allocation is on a 25 kHz channel basis, it is clear that whatever the modulation the actual message bandwidth Δf_m needs to be less than the channel assignment bandwidth Δf_a, e.g., if $\Delta f_a = 25$ kHz, then Δf_m is usually equal to 13 kHz.

To some extent it could be suggested one is 'wasting' much of the radio spectrum because of channelization. A way around the problem is of course to assign channels on a staggered space dispersed basis. In TDMA the users are grouped in fewer wider channels and the 'waste' problem is reduced. However power control, delayed multipath and a near-far problem can arise.

In digital modulation a reduced bit rate per Hertz gives the possibility of having more data (user) channels. Unfortunately for speech this is not easily achieved in practice, as described in Chapter 4B, and the necessary channel bandwidth remains a definite problem.

3.3   How bandwidth limits user numbers

Conventional land mobile usually implies a common base station, but now public mobile radiotelephony usually means cellular, CT2, etc, and in the future PCNs. It is possible to derive a simple guide formula for the number of radio subscribers, in a province or city, who can expect to be offered a service at any particular time. For example let

Area of the City being considered   =   A kms^2
Population of the City   =   P thousands of people
Avg. radius of a radio cell   =   R km
Number of radio channels in each cell   = n_c

n_c will depend on how the cell repeat pattern is organized and also the spectrum allocated to the service. In TACS, $n_c \sim 27$; in ETACS, $n_c \sim 57$, using a twenty-one cell repeat pattern, 25 kHz FM and the allocation plan Table 1.1. The cell plan can be varied, for example, see Fig. 3.2 which used to apply within the London M25 area.

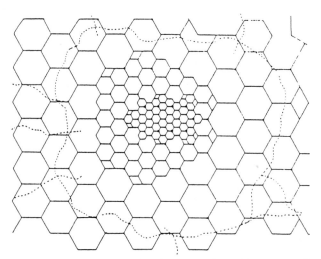

Fig. 3.2   A typical city radio cell plan - the cells are smaller where the most users are expected

A radiotelephone system designed on a cellular basis also uses a dynamic assignment or trunking technique (Chapter 6) which can be interpreted as a channel gain, i.e. typically 30 users think they each have individual access to a channel. Therefore the number of users supported by each radio channel in the cell appears to be multiplied by some factor.

We will use a factor of times 30; approximated as equal to 10π in the equation below.

$\therefore$ Number of users $\approx 30\, n_c \approx 10\,\pi n_c$

Meanwhile number of cells in city $= \dfrac{A}{\pi R^2}$ (uniform cells)

$\therefore$ Total number of users supported $\approx 10\,\pi n_c \times \dfrac{A}{\pi R^2}$

As a percentage of number of population $\approx \dfrac{10\, n_c A}{P \times 1000 \times R^2}\, 100$

or $\qquad \% = \dfrac{A n_c}{P R^2}$ $\qquad\qquad$ ...(3.1)

Example

Consider the City of Bristol

Here $P \approx 500{,}000 = 500$
$\quad\ A \approx 120\ \text{km}^2$
using $R = 2\ \text{km}$
and $\quad n_c = 40$

$\% \text{ offered service} = \dfrac{120 \times 40}{500 \times 4} = 2.5\% \text{ of the population}$

Eqn. 3.1 shows that highly populated cities (A/P ratio small) are the most difficult to serve, i.e. London; also the cell size is critical, whilst one must maximize the number of channels in the allocated radio spectrum.

Eqn. 3.1 can be re-written involving the total bandwidth ΔF allocated to the service, the bandwidth assigned or allocated to each user Δf_a, and cell cluster size N, since

$$n_c = \dfrac{\Delta F}{\Delta f_a N}$$

$$\% \; = \; \frac{A \, \Delta F}{P \, R^2 \, \Delta f_a N} \qquad\qquad ...(3.2)$$

This equation shows that apart from the global terms, the success of a cellular arrangement depends on the spectrum allocated per user, i.e. the term Δf_a, which as we saw above is greater than the signal bandwidth Δf_m.

Since Eqn. 3.2, as explained, is an approximation, because of the trunking gain factor and the population distribution factor which is unlikely to be uniform, plus also Δf_m equals a percentage of Δf_a, usually only $\approx 50\%$, this implies we can use Δf_m (the users signal modulation bandwidth) as the benchmark of *any system design*.

This matter of spectrum efficiency in a cell structured radio system is also discussed in detail by Calhoun [3], Hirade [4], and Lee [5]. They extend the discussion however by also considering co-channel interference which is a problem mainly centred around the cell layout geometry (N) and the co-channel interference threshold. 25 kHz analog FM is good in this respect because it achieves a significant demodulation gain [6], but in order not to lengthen this Chapter discussion on this matter is not considered further.

3.4   Analog modulation bandwidths

3.4.1 FM

Conventional FM will remain the mainstay of conventional PMR for some time to come. It is an established technology, and certainly 12.5 kHz FM compactly occupies the signal bandwidth Δf_m for the transmission of speech, although 25 kHz FM has noticeably superior 'sound quality' e.g. on TACS cellular.

The FM signal waveform can be written as

$$s(t)_{FM} \; = \; \cos (\omega_c + \beta m'(t))t \qquad\qquad ... (3.3)$$

where β =   modulation index

$$= \; \frac{\text{maximum frequency deviation}}{\text{modulation frequency}}$$

$$= \; \frac{\Delta f}{f_m}$$

To generate FM it is necessary to cause a frequency deviation of the transmitter output frequency, i.e. this being the so-called instantaneous carrier frequency f_i. In order to have a stable transmitter oscillator, it is essential to minimize its deviation, but if the oscillator is only a submultiple of the final output frequency, multiplication of the carrier can be used to generate the *desired frequency and correct FM index, more accurately.*

Analog FM theory is very well established, but it is useful to introduce certain features as a precursor to digital shift key modulation.

Fig. 3.3 shows how the carrier waveform changes in frequency, but is transmitted with a constant amplitude. Note also how the FM has a non-regular zero crossing, meaning the carrier changes phase over the modulation cycle.

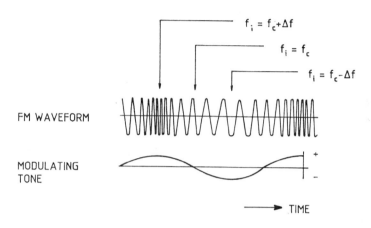

Fig. 3.3   The instantaneous frequency of an FM waveform in relation to sinusoidal modulation

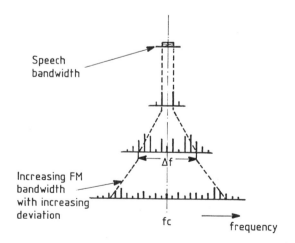

Fig. 3.4   The corresponding spectrum of an FM waveform showing how the number of sidebands (equal spacing) increases with the modulation index

The spectrum shown in Fig 3.4 widens as the deviation (Δf) increases and the FM package can be viewed as a group of symmetrically packed sidebands. The operation with a voice message can be gauged from Table 3.1 which is based on the definition of β, namely:

$$\beta = \frac{\Delta f}{f_m}$$

Numbers can be added as set out in Table 3.2.

Modulation change caused by speech	Effect on the index β	Effect on FM signal spectrum of the change
Increase in intensity (of tone)	Increases	Spectrum widens - more lines
Increase in tone (frequency)	Decreases	Same deviation - but less lines in same BW
Decrease in tone (frequency)	Increases	Same deviation - now more lines in same BW
Speech modulation Note -	High tones less intense than low tones	FM spectrum constantly occupies same BW

Table 3.1: Effect of a signalling tone and voice on the FM signal spectrum

FM index	Δf kHz	No. of Sidebands	Carson kHz	S/N improvement at receiver
	(1)	(2)	(3)	(4)
0.8	2.4	4	10.8	+ 5 dB
1.5	4.5	6	15.0	+ 11 dB

Table 3.2: Low index FM performance data

Notes: - (1)   Using $\Delta f = \beta.f_m$ where $f_m = 3\text{kHz}$
        (2)   Calculated from significant BW
        (3)   Using BW = $2\Delta f + 2f_m$
        (4)   According to demodulation improvement and pre-emphasis techniques [6]

3.4.2   DSB-SC

Double sideband reduced or suppressed carrier *amplitude modulation* (DSB-SC) is actually an example of a modulation where the signal bandwidth best fills the available bandwidth. Its signal-to-noise performance is as good as, if not better than 25 kHz FM, yet only occupies less than 6 kHz bandwidth and requires no in-band tones, etc [7]. It is introduced here, not as a recommendation for use in PMR or PCNs, but because the principle of generation leads directly into digital modulation.

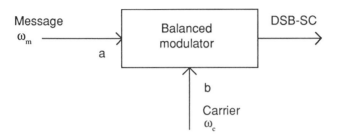

Fig. 3.5   Generation of DSB-SC using a balanced mixer
and showing dc offset components

Thus, Fig. 3.5 shows how DSB-SC is generated using a double-balanced modulator, or multiplier, such as the Plessey Semiconductor device type SL 640. In practice complete balance of the signal input ports is not achieved, and the equation of operation of a practical modulator, including the dc offset components a and b, should be written as

$$v_o(t) \quad = \quad v_m(t) \; . \; v_c(t)$$

$$= \quad (v_m \sin \omega_m t + a) \; . \; (v_c \sin \omega_c t + b)$$

$$= \quad v_m.v_c \sin\omega_m t \sin \omega_c t + a.b + a.v_c\sin\omega_c t + b.v_c\cos\omega_m t \qquad ... (3.4)$$

The desired output (the product) is the first term in Eqn. 3.4. Mixer unbalance and any non-linearity will cause difficulty with practical modulator realization, especially when the output of two quadrature mixers are added together as described below.

The waveform of DSB-SC is partly shown in Fig. 3.7(e) below when describing digital phase shift keying modulation. The fully filtered PSK waveform will have its sidebands contained as a single pair when a single on/off waveform is considered.

An actual example of a two-tone DSB-SC spectrum is shown in Fig. 3.6. This RF signal at 318 MHz is directly generated from baseband using an active FET mixer. The purpose of showing the diagram, however, is to draw attention to the limitation of practical RF technology in terms of in- and out-of-band intermodulation components and the non completely suppressed carrier due to the mixer offset components. These and other limitations make the realization of RF modulated signals more difficult than one might hope for.

Linear SSB generation requires repetition of the modulation process and other circuits in order to remove one of the sideband components. SSB offers 5 kHz BW channel spacing for mobile speech circuits in theory; a full description and overview of transparent tone-in-band SSB modulation has recently been given by Bateman [8].

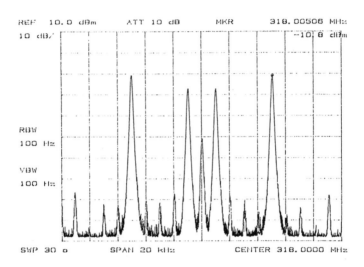

Fig. 3.6   Practical two-tone DSB-SC signal at 318 MHz produced by direct product modulation

3.5 Shift key modulations

Digital modulation leads to a family of what are known as shift key modulations. Let us assume the binary data is coded in some suitable form such that:

a '1' or mark waveform = +1

over the symbol period T

a '0' or space waveform = -1

3.5.1 PSK

Applying this signal to a balanced modulator of Fig. 3.5, with the amendments shown in Fig.3.7(a), leads to binary PSK or bipolar ASK (BPSK), i.e.

$$s(t) = \pm \cos \omega_c t \qquad \qquad ...(3.5)$$

The output waveform shown in Fig. 3.7 (d) has a constant amplitude, and a phase change of 180° at each mark/space transition. The spectrum of this waveform extends well beyond the symbol* rate frequency offset either side from the (suppressed) centre carrier, i.e.,

$f_r = 1/T$ either side of the suppressed carrier frequency f_c.

To contain the spectrum, shaping or filtering of the data symbols has to be employed, i.e. Fig. 3.7 (e).

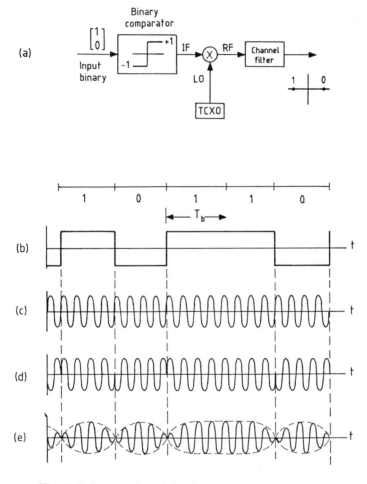

Fig. 3.7 Generation of the binary PSK signal
(a) modulator, (b) data waveform, (c) carrier
(d) unfiltered PSK, (e) filtered PSK

---

* In multi-level modulation the bit rate exceeds the symbol rate by $\log_2 L$, where L = number of signalling levels.

The ideal binary Nyquist transmission rate in a bandpass channel of bandwidth B (for example see Feher [4]), is:

$$f_r = B$$

Vestigial shaping of the channel filter would most likely be employed, which reduces f_r to the value:

$$f_r = \frac{B}{1+\alpha} \qquad \qquad ...(3.6)$$

where $0 < \alpha < 1$

        (ideal)   (cosine squared)

A change in envelope amplitude of the signal i.e. Fig. 3.7(e), is seen to occur at the mark/space transition, and therefore more complex quadrature modulation is used, as described below.

3.5.2   FSK

FSK implies that one switches from a mark frequency to a space frequency, in sympathy with the data, i.e. Fig. 3.8.

Carrier oscillators

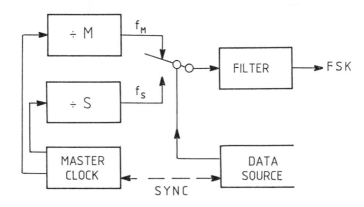

Fig. 3.8   A means of generating FSK

The spectrum of the waveform

$$s(t) = \cos (\omega_c \pm \Delta\omega) t \qquad \qquad ...(3.7)$$

can be calculated in the case of a regular 1,0,1,0 pattern, by adding the spectrum of the symbol rate time limited carrier waveforms:

f_m = f_c - Δf for period T

f_s = f_c + Δf for period T

One finds the signal energy concentrates at the mark and space frequencies, i.e. Fig. 3.9. This type of signal is used in paging applications (Chapter 7).

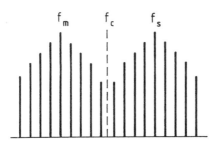

Fig. 3.9 FSK signal spectrum used in paging system, m = 18

The modulation index of FSK is given by

$$m = (f_s - f_m).T = \frac{f_s - f_m}{f_r} \qquad ...(3.8)$$

In Fig. 3.9, m = 18. To increase the data transmission rate, but not the bandwidth occupied, low index FSK schemes must be employed. A common standard is fast frequency shift keying FFSK [7, 9] where

f_m	= 1200 Hz	note the nominal carrier
f_s	= 1800 Hz	frequency f_c = 1500 Hz
f_b	= 1200 b/s, = symbol rate frequency f_r in this case	

$$\therefore m = \frac{1800 - 1200}{1200} = 0.5$$

Here one cycle of 1200 Hz is followed by one and a half cycles of 1800 Hz, with no phase discontinuity at the bit interval. The spectrum is now concentrated in the band 600 to 2400 Hz.

It can be shown [11,12] that FFSK is equivalent to minimum shift keying (MSK), a particular form of quadrature phase shift keying (QPSK), which is described below. With all three of these modulations however the modulation signal has significant adjacent channel sideband components, which make them unsuitable for future digital cellular radio systems where good adjacent channel performance is an essential requirement.

The solution to the difficulty is to shape the binary signal waveform by suitable wave filtering. In the case of FSK, such as Fig. 3.8, the conventional method is to constrain the change over from a mark to a space etc with a filter in the data line of Fig. 3.8.

The coherence between the mark and space tone are kept by arranging

$$f_m \quad = \quad f_c - f_b/2$$

$$f_s \quad = \quad f_c + f_b/2$$

where f_b =   bit rate

The modulation index is now equal to 1 which implies the in-band spectrum is fuller. The method has been described as carrier frequency exchange keying (CFEK) [10]. The distant adjacent channel spectrum is superior to MSK, and non-coherent FSK.

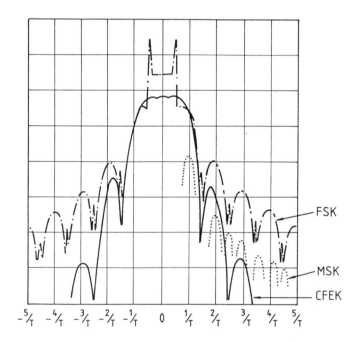

Fig. 3.10   The adjacent and in-band spectrum occupancy of
FSK (m = 1), MSK (m = 0.5) and CFEK (m = 1)

3.5.2.1 *Multi-level FSK*

The data throughput can be increased by using more than two frequencies. Thus using L levels, or frequencies, a symbol can represent $\log_2 L$ bits, i.e. 4 frequencies will convey 2 bits per symbol. This technology is being explored for the new European Radio Messaging System (ERMES) - see Chapters 7 and 15.

The data rate required is 6.25 kb/s; thus requiring a symbol rate of 3.125 kb/s, and use the four signalling frequencies [13]

$$
\begin{aligned}
f_{00} &= f_c - 3f_b/2; &\text{offset} &= -4687.5 \text{ Hz} \\
f_{10} &= f_c - f_b/2; &\text{offset} &= -1562.5 \text{ Hz} \\
f_{11} &= f_c + f_b/2; &\text{offset} &= +1562.5 \text{ Hz} \\
f_{01} &= f_c + 3f_b/2; &\text{offset} &= +4687.5 \text{ Hz}
\end{aligned}
$$

The arrangement one notes is a four-level equivalent of the two-level CFEK and m = 1.0; at least between adjacent levels.

The data is shaped using a 10th order Bessel filter, with a 3dB point at 4 kHz, so that the carrier oscillator can move smoothly from one signalling frequency to the next within the symbol period, i.e. 1/3.125 msec, as sketched in Fig. 3.11. It is claimed that adjacent channel interference levels of better than -70 dB can be achieved [13]. A three level conventional limiter-descriminator detector can be used provided the signal receiving conditions are good.

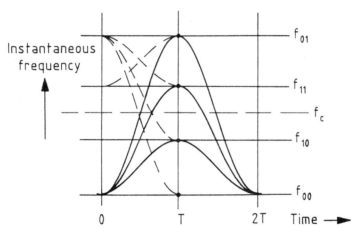

Fig. 3.11   The possible instantaneous frequency positions
of 4 level FSK over a symbol period

3.5.3 Quadrature PSK

As explained for FSK, so four phase levels will provide two data bits per symbol for PSK. As a result quadrature PSK (QPSK) provides twice the data throughput in the same bandwidth when compared to BPSK. On the other hand QPSK needs two 90° offset multipliers as shown in Fig. 3.12.

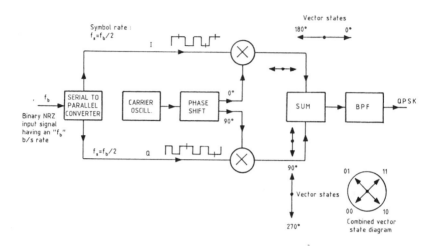

Fig. 3.12   Conventional method of generating QPSK

The theory of QPSK is dealt with all text books on digital modulation and leads to the idea of the basic signal phase constellation diagram shown firstly an an insert in Fig. 3.12. This combined vector state diagram shows where the signal phase is *at the centre of each symbol period.*

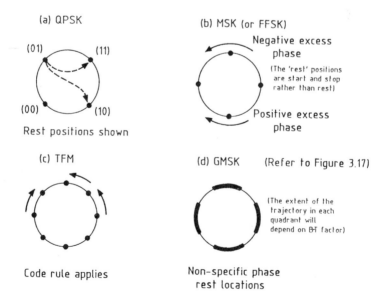

Fig. 3.13   The movement of the carrier phasor during digital modulation signalling for the four cases QPSK, MSK, TFM and GMSK

The interesting question is what happens as one moves from one symbol (say 01) to the next (say 11). If there is no coherence between the data rate and the carrier frequency the waveform may well be constrained to move across the phase point circle as indicated, or rapidly round the circle, shown again now in Fig. 3.13. Pasupathy [12] describes the solution to the problem in his description of MSK modulation. By combining off-set PSK in the lower path Q of the QPSK modulator Fig. 3.12 with shaping of the data pulses so they follow a half sine waveform profile, the modified QPSK waveform, now called MSK, acquires a constant envelope and continuous phase waveform, in fact equivalent to FFSK.

3.5.3.1 *Filtered MSK*
Minimum shift keying achieves the objective of having the carrier move from one phase state to the next, around the phase circle as shown in Fig. 3.13 (b) in $\pi/2$ increments. To do this the frequency must either be ahead of the nominal carrier frequency, or behind, i.e. either f_s or f_m. As for FFSK the shift of the carrier frequency is $\pm$ 300 Hz i.e., one quarter the bit rate of FFSK. The so-called frequency deviation f_s to f_m (600 Hz) is equal to half the bit rate (1200 b/s) giving m = 0.5. However MSK still suffers from poor adjacent channel sidebands due to its FSK type spectrum.

Amorosa [14] first described how constraining this change of frequency by filtering could improve the situation. The diagram Fig. 3.14, adapted from De Jager and Dekker [15] demonstrate the point. The sudden switch in frequency from say f_m to f_s causing the phase to move by $\pi/2$ radians in a symbol period is smoothed out and some improvement on the out-of-channel power spectral density results, which are illustrated in Fig. 3.15, occur but not significantly.

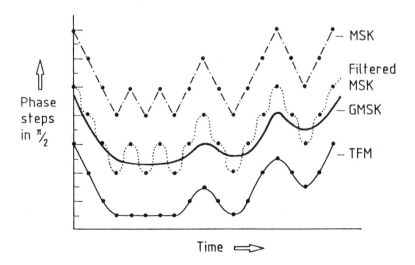

Fig. 3.14   The phase movement and accumulation as a function
of the data for MSK, filtered MSK, TFM and GMSK

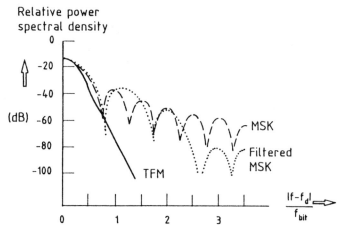

Fig. 3.15   The adjacent channel power spectral density of MSK, filtered MSK and TFM

For much greater improvement, but still retaining the important bandwidth efficient constant envelope property two approaches have been developed.

3.5.4   Tamed FM (TFM)

The approach made by De Jager and Dekker [15] was to correlatively code the data stream so that less and smaller phase changes take place. The code rule they proposed establishes that a change of phase by $\pi/2$ only takes place if the three succeeding bits have the same polarity; no phase change takes place if the three bits are of alternating polarity. Polarity changes of $\pi/4$ are reserved for the bit configurations 110, 100, 011 and 001. The result is a much more constrained movement around the phase circle as depicted in Fig. 3.13(c) with a correspondingly much improved spectral performance. It must be recognised however that the phase 'rest' positions are now placed at $\pi/4$ angles around the phase circuit which must be taken into account in the receiver demodulator design.

3.5.5   Gaussian minimum shift keying (GMSK)

Sunde [16] in his papers on the fundamentals of pulse transmission drew attention to the fact that whereas the classic Nyquist raised-cosine filter is completely free of intersymbol interference (ISI), a Gaussian shaped filter, Fig. 3.16, having the envelope shape,

$$A(\omega) = \exp^{-0.54} \left(\frac{\omega}{\omega_c} \right) \qquad \qquad ...(3.9)$$

produces a response which has only 1% ISI, but considerably better adjacent symbol performance.

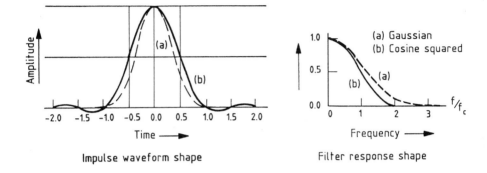

Impulse waveform shape

Filter response shape

Fig. 3.16   The impulse response of (a) a Gaussian low pass filter
and (b) a cosine squared amplitude shaped LPF

Gaussian pulse pre-shaping was added to MSK by Murota and Hirade [17] who
termed the modulation Gaussian MSK (GMSK). The pulse shape of a Gaussian
filter is also Gaussian, hence the ISI.   One has a choice between the premodulation filter bandwidth B and the bit period T. If B > 1/T, then the waveform
is essentially MSK; if however B < 1/T, the change of symbol cannot reach its next
position in the time allotted. The effect is shown in Fig. 3.14, when B < 1/T. The
constellation becomes blurred due to the influence of remnants of previous pulses
on the phase change, i.e. Fig. 3.13 (d). The phase behaviour attempts to follow
Amorosa's positions but is constrained.   In the receiver a complementary
Gaussian filter will recover the data because it will again be influenced by the
pre-designed intersymbol interference.   It does mean however that the phase
modulation pattern must be generated accurately otherwise the pre-assigned
construction cannot be interpreted. A BT product = 0.3 is planned for GSM and
a discussion of the design of synthesizers for GMSK working at ~ 270 kb/s has
been published [18].

It is also worth noting the so-called trellis diagram of these phase shift keyed
modulations, shown in Fig. 3.17.

In the case of MSK phase changes of π/2 occur at each data interval. All 1's
cause the phase to move upwards at 45°; all 0's downwards at -45°. Other patterns
will form the trellis.

In the case of GMSK, depending on the BT product value, the phase change
is unable to follow the MSK trellis. This is why in the GSM signalling protocol
a burst data mode is necessary in order to maintain a phase reference which
would become uncertain after a long string of 1's or 0's.

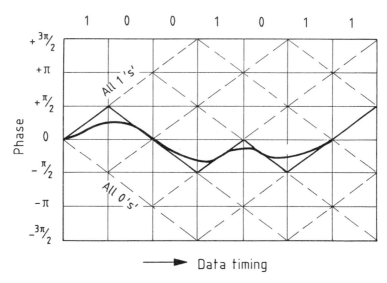

Fig. 3.17   The trellis diagram of the carrier phase positions formed
by data changes during MSK and GMSK modulation

Measurements confirm that the power spectral density of GMSK does indeed
exceed that of MSK in respect adjacent channel performance and a 60 dB out-of-
band performance ratio can be achieved. GMSK (modulation index = 0.3) is in
theory a constant amplitude waveform; unfortunately the fact that the carrier
must be power controlled and also confined to timed bursts implies that the
overall carrier envelope will cause some spectral spreading as well as denying the
use of Class-C transmitter stages. The spectrum spreading problem has been
studied by O'Reilly and Kokkos [19].

3.6   Bit error rate

Digital modulation does not suffer an apparent worsening S/N ratio as does
analog modulation; what happens is digits become mis-interpreted and an
increasing bit error rate (BER) response sets in. The probability of an error being
recorded versus the ratio of the energy per bit E_b to the noise power per Hertz N_o
can be calculated [6] and for QPSK is:

$$P_e = \frac{1}{2} \, \text{erfc} \left(\frac{E_b}{N_o} \right)^{\frac{1}{2}} \qquad \qquad ...(3.10)$$

where $\dfrac{E_b}{N_o} = \dfrac{C \, B}{N \, R}$ where C/N $= \dfrac{\text{carrier}}{\text{noise}}$ ratio   ...(3.11)

Eqn. 3.10 gives a result shown marked theoretical in Fig. 3.18. In practice receiver and transmitter RF circuits are added to the data transmission system and also a multipath will exist in the propagation environment. The observed BER worsens, as shown, and is often limited to some irreducible level. These three cases are marked laboratory, static and moving at 30 mph in Fig. 3.18, respectively [20].

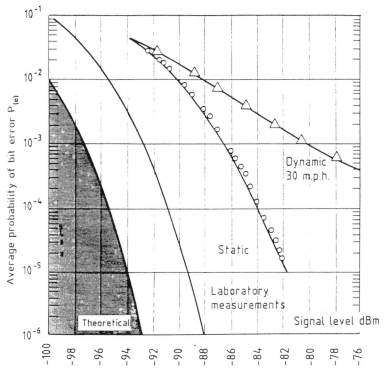

Fig. 3.18   Expected and observed bit error rate performance for 2400 b/s MSK over a 12.5 kHz VHF radio channel at 172 MHz

The causes of static imperfections have been studied in detail [21] and are listed below:

1   Modem imperfections - IF back-to-back:

1.1   Phase and amplitude errors of the modulator; see para 3.4.2.
1.2   Intersymbol interference caused by the filters in a back-to-back modem
1.3   Carrier recovery phase noise
1.4   Differential encoding/decoding
1.5   Jitter (imperfect sampling instants)
1.6   Excess noise bandwidth of receiver (demodulator)
1.7   Other hardware impairments (temperature variations, aging, etc.)

2   RF channel imperfections:

2.1   AM/PM conversion of the quasi-linear output stage
2.2   Band-limitation and channel group delay
2.3   Adjacent RF channel interference
2.4   Feeder and echo distortion

These imperfections can easily amount to 6dB or more, i.e. carrier-to-noise needs to be this much higher to achieve same BER.

If the digital modulation schemes forms part of a mobile system, multipath propagation causes further degradation in the system performance and necessitates channel sounding and path equilisation techniques, i.e. Chapters 2 and 14

3.6.1   Improving BER

Data errors can be reduced, or corrected by Forward Error Correction (FEC) techniques (see Chapter 4). Unfortunately FEC implies transmitting more data, which means either (i) accepting more errors, or (ii) lowering the data transmission rate, or (iii) increasing the signal power. The last implication is contained in Shannon's formula

$$\frac{R}{B} < \log_2 \left[1 + \frac{E_b\, R}{N_o\, B} \right] \qquad ...(3.12)$$

which says that if the bandwidth is limited, one can only increase the transmission rate provided the minimum E_b/N_o ratio increases faster.

The way around the conflict is to combine the coding and modulation [22]. The data message is now pre-coded before modulation in specific patterns, such as a set-partioned code [23, 24], and this code can be recognised with added confidence in the received signal. Hence use of the code saves transmitter power on a per-data bit basis without increasing the signal bandwidth. *Coding gains* of up to 6 dB in E_b/N_o ratio have been predicted. The use of so-called *trellis coding* in the GSM system implementation is described in Chapter 14. A universal approach to trellis-coded modulation (TCM) is described in reference [25].

3.7   Universal modulators

It may have been noticed that all forms of modulation described hinged on the use of one or two balanced modulators, in fact, in the ideal case the digital coded or processed signal can be modulated directly to RF. This has led to the concept of the generalized modulator structure [26] shown in Fig. 3.19, based on the equation of operation:

$$s(t) = a(t)\,.\,\cos\omega_c t + b(t)\,.\,\sin\omega_c t \qquad ...(3.13)$$

In analog modulation, a(t), or b(t), or both are continuous. In digital modulation a(t), or b(t), or both are set or partially set over each symbol period, by the precursory digital signal processing (DSP) circuits. In the case of coded modulation the DSP would also contain memory.

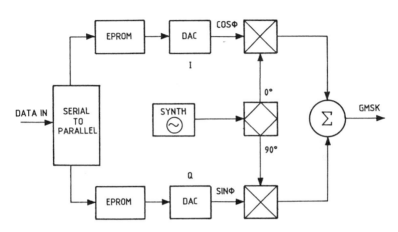

Fig. 3.19   Universal I/Q modulator structure

Because most digital modulations require a continuous phase trajectory of the carrier, the same modulation can also be generated by direct phase modulation of the carrier, i.e. Fig. 3.20. [18] The direct synthesis of a microwave signal from baseband, as implied in Figs. 3.19 and 3.20, can put excessive demands on the synthesizer design and several other arrangements exist. The demodulation of these signals, which is not dealt with in this Chapter, requires a similar, but reverse ordered complex processing. Carrier synchronization is also a technical matter which must be faced in coherent demodulation circuits;   non-coherent demodulation on the other hand avoids this added complexity but with some performance penalty.

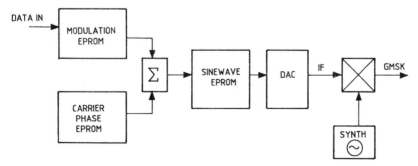

Fig. 3.20   Universal phase/frequency modulator structure

3.8   References

1.   CARSON, J. R.: 'Notes on the theory of modulation', Proc. IRE, 1922, 10, Feb, p. 57

2.   ARMSTRONG, E. H.: 'A method of reducing disturbances in radio signalling by a system of frequency modulation', Proc. IRE, 1936, 24, May, p. 140

3.   CALHOUN, G.: 'Digital cellular radio' (Artech House, New York, 1988)

4.   FEHER, K. (ed): 'Advanced digital communications' (Prentice-Hall, New York, 1987)

5.   LEE, W. C. Y.: 'Spectrum efficiency in cellular', IEEE Trans. 1989, VT-38, May, p. 69

6.   COUCH, L. W.: 'Digital and analog communication systems' (Macmillan, New York, 1983)

7.   HOLBECHE, R.J. (Ed): 'Land mobile radio systems' (Peter Peregrinus, London, 1985), Ch.6 and Ch. 7

8.   BATEMAN, A.J.: 'Feed forward transparent tone-in-band - its implementation and applications', IEEE Trans. 1990,VT 39, p. 235

9.   MPT Code of Practice 1317: 'Transmission of digital information over land mobile radio systems' (HMSO 1981)

10.   HARRIS, R. M.: 'Coherent frequency exchange keying (CFEK)', Royal Aerospace Est., Report 8808, 1988, February

11.   ZIEMER, R.E., and TRANTER, W.H.: 'Principles of communications, systems, modulation and noise' (Houghton Mifflin, Boston, 1990)

12.   PASUPATHY, S.: 'Minimum shift keying: a spectrally efficient modulation', IEEE Comms Magazine, 1979, July, p.14

13.   YARWOOD, A., HOLMES, W.H., and GIDLOW, A.C.: 'European radiopaging', British Telecom Tech. J., 1990, 8, Jan, p.1

14.   AMOROSO, F.: 'Pulse and spectrum manipulation in the minimum frequency shift keying format', IEEE Trans., 1976, COM-24, March, p. 381

15.   de JAGER, F., and DEKKER, C.B.: 'Tamed frequency modulation, a novel method to achieve spectrum economy in digital transmission', IEEE Trans., 1978, COM-26, May, p.534

16. SUNDE, E. D.: 'Theoretical fundamentals of pulse transmission', BSTJ, 1954, 33, May, p. 721 and Pt II, July, p. 987

17. MUROTA, K., and HIRADE, K.: 'GMSK modulation for digital mobile radio telephones', IEEE Trans., 1981, COM-29, No. 7, July, p. 1044

18. MEYERS, R.A., and WATERS, P.H.: 'Synthesiser review for pan-European digital cellular radio', IEE Colloquium on VLSI for Mobile Telecomms Systems, 14 March 1990

19. O'REILLY, J. J., and KOKKOS, A.: 'Spectral considerations for time division duplex CPM with application to future personal communications', IEE Conf. Mobile Radio, 1989, Warwick, No. 315, December, p. 137

20. EL-JAAFREH, Y.G.D., and MACARIO, R.C.V.: 'Experience with a multirate modem system for data transmission over 12.5 kHz land mobile radio', IEE Colloquium on Multi-level modulation techniques; point-to-point and mobile radio, London, 19 March 1990

21. COMPSTON, I.K.: 'Relating constellation parameters to digital radio performance', Hewlett-Packard Publication No. 5954-7943

22. IEEE J. Selected Area Comms, 1989, 'Bandwidth and power efficient coded modulation', Aug. and Dec. issues

23. UNGERBOECK, G.: 'Channel coding with multilevel/phase signals', IEEE Trans. Info. Theory, 1982, IT-28, January, p. 55

24. CLARK, G.C., and CAIN, J.T.: 'Error-correction coding for digital communications' (Plenum Press, New York, 1981)

25. VITERBI, A.J., WOLF, J. K., ZEHAVI, E., and PADOVANI, R.: 'A pragmatic approach to trellis-coded modulation', IEEE Comms. Magazine, 1989, July, p.11

26. DAVARIAN, F., and SUMIDA, J. T.: 'A multipurpose digital modulation', IEEE Comms. Magazine, 1989, February, p.36

Chapter 4

Signal coding A: Channel coding

Professor Paddy Farrell

4.1 Introduction

This chapter is concerned with some of the basic coding processes required in order to transmit digital signals over mobile radio channels. The various processes involved are set out in Fig. 4.1. The choice of a modulation, coding or multiple access method is determined by the need to achieve communication efficiency and adequate performance under the conditions imposed by the radio transmission path.

In this context it is relevant to recall Shannon's equation [1] relating the information capacity, C_t (in bit/s) of a channel to its bandwidth, B, and the signal-to-noise power ratio, S/N

$$C_t = B \log_2 [\ 1 + S/N \ ]$$

This assumes additive, white, Gaussian noise (AWGN) on the channel, so it is only an upper bound to the bit rate achievable on an actual mobile radio channel, which is subject to fading, interference, and other impairments. In addition, achieving rates close to capacity implies very complex, and therefore expensive, coding circuits. Thus, in practice, it is necessary to limit the bit rate on the mobile radio channel, and to use relatively simple but effective processing and coding techniques to ensure that the digital signal is conveyed reliably.

Four relevant processing and coding techniques are introduced in this chapter. Section A covers channel coding, which may be defined as error detection and correction (EDC) coding, for protection against errors caused by noise or interference on the channel, and transmission (line) coding, which matches the signal to the characteristics of the channel. Security (cryptographic) coding, which prevents unauthorised access to the signal, will also be mentioned, but the main emphasis will be on EDC coding. Section B is devoted to speech processing and coding.

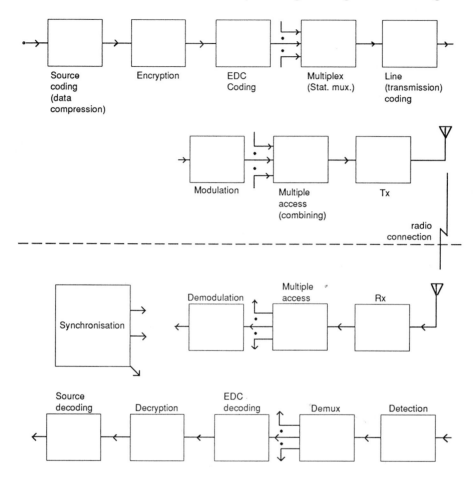

Fig. 4.1 Coding and modulation processes

4.2 Transmission coding

One advantage of digitally coded signal formats is the ease with which they can be further coded in order to modify their spectral or temporal characteristics. Spectral nulls at DC and the baud rate can be inserted, for example (e.g. *alternate mark inversion* (AMI), *pairselected ternary* (PST), *split-phase* or *Manchester* coding); spectral line components can be enhanced so as to facilitate synchronisation (e.g. *alternate digit inversion* (ADI), *return-to-zero* (RZ) coding); and bandwidth can be compressed (e.g. *Miller coding* or *delay modulation*). These and other similar coding and scrambling techniques are all forms of *line, channel* or *modulation* coding [2].

Manchester coding, for example, was invented in the Department of Electrical Engineering at Manchester University, in a power transmission context. It is also called *biphase* (BIØ) or *split-phase* coding. Each binary digit is replaced by a pair of binary digits:-

$$0 \rightarrow 1\ 0$$

$$1 \rightarrow 0\ 1$$

The basic waveforms of Manchester coding are first-order Walsh functions, so Manchester coding is also called WAL-1. A modification of Manchester coding is thus to use WAL-2 waveforms, where

$$0 \rightarrow 0\ 1\ 1\ 0$$

$$1 \rightarrow 1\ 0\ 0\ 1$$

which is very tolerant to channel distortion. Manchester coding is used in the TACS cellular system (e.g. see [3]).

Transmission coding methods almost always put some redundancy into the signal (e.g. Manchester coding doubles the bit rate). This redundancy can also be used for EDC purposes. AMI violations indicate the presence of errors, for example. It is also possible and effective to combine transmission and EDC coding [4].

4.3   Security coding

Two reasons for using cryptographic techniques [5] in mobile communications are (i) privacy, to prevent unauthorised persons from extracting information from the channel (e.g. 'wire-tapping'), and (ii) authentication, to prevent unauthorised persons from injecting information into the channel (either for their own use or to disrupt or confuse legitimate communication). The message, or *plaintext*, p, is enciphered (or encrypted) with an invertible transformation, E_k, to produce the *ciphertext*, $c = E_k(p)$. The ciphertext is transmitted over the public (i.e. basically insecure) radio channel. When c is received by an authorised receiver, it is deciphered (or decrypted) using the inverse transformation, E_k^{-1}, to obtain the original plaintext:-

$$E_k^{-1}\ (c) = E_k^{-1}\ (E_k(p)) = p\ .$$

E_k is chosen from a family of cryptographic transformations, often regarded as public information. The parameter k represents the *key*, which selects the individual transformation out of the family to be used in a given transmission, and is safeguarded. The key to be used is communicated to the community of authorised users over a private, secure channel (e.g. by courier!).

Encryption schemes fall into two broad categories: *block encryption*, in which the plaintext is segmented into blocks of fixed size, each block being enciphered independently of the others, such that a particular plaintext block is enciphered into the same ciphertext block each time it appears in a message; and *stream encryption*, in which the plaintext is enciphered continuously, in such a way that consecutive symbols are not treated independently, and two occurrences of the same plaintext input are unlikely to be enciphered into the same ciphertext.

The basic operations performed in block encryption schemes are *permutation* (re-arrangement) and *substitution*. Permutation is a *linear* operation, and has one major drawback when used alone: it is vulnerable to 'trick' messages, which may identify the permutation being used. Substitution transformations, on the other hand, are *non-linear*. The number of different substitution (or connection) patterns possible with blocks of n data bits is $(2^n)!$ which grows very rapidly with n. This makes a cryptanalyst's task (finding the key, i.e. the particular substitution used) difficult and, for large n, computationally impractical. However, it is also impractical to implement such a system, because of the large number (2^n) of connections required. Typically, several stages of permutation and substitution are interleaved to provide a more powerful cryptosystem (product cypher) than would be feasible using either technique on its own. In 1977, the American National Bureau of Standards adopted such a scheme, involving 18 stages of permutation and substitution, as their national *Data Encryption Standard* (DES). The enciphering and deciphering circuits for the DES can easily be implemented using conventional LSI technology [5].

In contrast to the above block encryption techniques, *stream encryption* uses a 'running key', which is combined with the data symbols in a continuous manner. The running key is usually a 'random' sequence of symbols drawn from the same set as that used to represent the plaintext data. The sequence symbols are often added to the data symbols (modulo 2 in the binary case) to produce the ciphertext. The random nature of the running key ensures that two occurrences of a given plaintext symbol do not generally result in the same ciphertext. If the running key is a truly random sequence of equiprobable symbols, of infinite length, a cryptanalyst would require an infinite time to 'crack' such a data-stream encryption scheme. The cryptosystem would therefore be *unconditionally secure*. However, to be practical, the running key must be of finite length, and is often a deterministic, periodic sequence, generated using a shift-register with non-linear feedback. The result is a system with sufficient but limited security, which can be 'cracked' only in a long, though finite, time.

In the *public key* cryptosystems mentioned earlier, the enciphering and deciphering functions are separated, in such a way that one is apparently (to an 'eavesdropper') independent of the other. Given any particular enciphering function, it is not feasible to deduce the corresponding deciphering function. In such a system, a public directory would indicate the *en*ciphering key to be used for each individual address, but the corresponding *de*ciphering key would be private to each address. The RSA (Rivest, Shamir and Adelman [9]) algorithm is one such public key scheme [5].

Three points are worth noting with regard to the use of security coding. Encryption can (and indeed should, for best performance) be provided without the need for additional redundancy, so that there is no communications penalty in using it. The degree of encryption required depends on the nature of the message and the threat to it, as determined by the user. Because of this, it is best if encryption is provided at user terminals. However, it is important to provide at least sufficient privacy on mobile radio channels to avoid casual eavesdropping.

4.4   Error detection and correction coding

An EDC or *error control coding* [6-9] encoder protects digital information against errors by inserting *redundant* digits into the stream of digits to be transmitted. The redundant digits are calculated from the information digits (e.g. by *parity checking* or *modulo-2 addition*); this dependence between the two sets of digits is used in the decoder to detect and correct errors (e.g. failure of a parity check indicates a single error in one of the digits concerned). The process of encoding is one in which blocks of k input (information) digits are transformed into output blocks of n coded digits. The coded digits are almost always *binary* in practice, though a *multilevel* exception will be noted below (multilevel digits can always be coded into binary form for transmission). If only one input block is encoded at a time, then a *block code* results; decoding need only be concerned with single blocks (the n-digit *block length*) at a time.   If   several input blocks (h, say) determine a given output block, then a *tree code* is formed, and decoding extends over at least h blocks or more.   In all cases of practical importance, the encoder performs a linear computation on the input digits (i.e. in the binary case, the exclusive-OR, parity check, or modulo-2 addition operation). So all useful codes are *linear codes*, and *non-linear codes* are only of theoretical interest. The reason for this is that the linearity confers a high degree of mathematical structure on the code, which can be exploited in the decoder, making it much simpler to implement. A linear tree code is a *convolutional code*. The *rate* of a linear code is defined as $R=k/n$;   a high rate code has a small proportion of check to information digits and vice versa. A high rate code is efficient in the sense that little channel capacity is lost to redundant digits; however, for a given n, its error control power will be less than that of a lower rate code. In practice, the useful range of R is from 1/2 to near unity.

The random error control power of a block code depends on the minimum number of positions in which any pair of n-digit encoded blocks (code words) differ. This number is called the *minimum* or *Hamming distance* of the code. For example, if all pairs of code words differ in at least 3 positions, then it is capable of correcting any single error in a code word.   In general, a code can correct t = $\lfloor(d-1)/2\rfloor$ errors, where d is the Hamming distance, and $\lfloor x \rfloor$ is the largest integer ≤x.   Codes can also be used to detect errors, up to e, given by e = d - 1;   or can be used to both correct some and detect more errors, in which case d = t + e + 1, where t ≤ e.

Similar relationships apply to convolutional codes. In this case, the distance is defined over either h blocks, the constraint length of the code (hn digits), or over an unlimited number of blocks: d_{min} or d_{free}, respectively; the appropriate measure depending on the decoding method to be used. If the information (input) digits appear explicitly in the encoder output blocks then the code is *systematic*; it is *non-systematic* otherwise. The advantage of a systematic code is that a first estimate of the information digits can be obtained by inspection. In the case of a linear block code, there is no disadvantage in being systematic, in terms of Hamming distance; the d_{free} of a non-systematic convolutional code can be up to twice that of a systematic code with the same parameters, however.

Decoding is a much more complex process than encoding, particularly for relatively long and high rate codes. The most general optimum form of decoding algorithm, for example, computes the Hamming distance between the received code word (or sequence of blocks, in the case of a convolutional code) and all possible received blocks or sequences (there will be 2^k of these for a block code, and $\geq 2^{hk}$ for a convolutional code), and then selects the word or sequence with smallest distance as the one most likely to have been transmitted (*minimum distance decoding*). For even moderate values of k (or hk) this is a very complex algorithm. Linearity helps to reduce the decoding complexity considerably (*syndrome decoding*), but on its own is not enough to make block codes practical. Additional structure is required, and this is provided by three general classes of block codes; *cyclic codes* (e.g. Hamming, BCH and Reed-Solomon codes), *array codes* (also called product, iterated or multidimensional codes), and *Reed-Muller codes*. Practical and efficient decoders are possible for all these codes, which exist for a wide range of parameters. The tree structure of convolutional codes make them practical to decode, by means of the *Viterbi algorithm*, for constraint lengths (h) up to 10. Longer constraint length codes can be decoded by means of *sequential decoding* algorithms, but these are complex and suffer from computational overflow problems. Certain convolutional codes can be decoded by very simple *threshold decoding* methods, which also apply to block codes (*majority logic decoding*); the penalty is poorer performance compared to minimum distance decoding. A particularly useful class of threshold decodable convolutional codes is the *diffuse codes*, which are effective against a mixture of burst and random errors [6].

An error-control code detects and corrects errors. This either improves the output error rate of a given channel, or permits a reduction in the S/N needed to obtain a desired output error rate. On the other hand, use of a code means that some energy must be shared between information and redundant digits. An error-control code is efficient if, for a given error rate, the S/N advantage exceeds the redundancy loss. This overall advantage is called the *coding gain* of the code; for an AWGN channel at low error rates, it is given approximately by (in dB) G = $10 \log_{10}(Rt)$ where t is the number of errors the block or convolutional code can correct; Rt is the code *quality factor*. Thus good codes have both high R and large t (which implies large d and n or hn). A famous theoretical result, Shannon's *coding theorem* [1], states that if the block or constraint length of a code is allowed to grow without limit, then the output error rate can be as close to zero, and the code rate as close to capacity, as desired. In practice, the longer the code the more complex it is to implement (especially to decode), but coding gains of up to 4dB are quite feasible, and up to 6 or 7dB not impossible.

Up to this point, *forward error correction* (FEC) has been assumed: the code is used to correct errors at the receiver, with no help from re-transmissions or other system protocols. An alternative scheme is to use the code to detect errors, and to automatically request a repeat (ARQ) of any blocks containing errors. Error detection is much easier to implement than error correction; however, an ARQ system requires a feedback channel and suitable buffering, and the decoding delay is variable and very dependent on the loop delay of the system. Some of the difficulties are overcome by using hybrid FEC-ARQ schemes [10].

In mobile radio systems transmission errors do not occur at random (as for AWGN channels), but occur in clusters or *bursts*, due to fading, interference, or use of certain modulation schemes (e.g. DPSK). There are codes which can control bursts of errors, but a practical alternative, often used, is to *interleave* the digits of a random EDC code, so that a single burst does not affect more than a few digits in each block of the code, after de-interleaving at the receiver. Unfortunately, this is not as efficient as using a code properly matched to the channel error pattern, but it has the advantage of being less sensitive to varying channel error statistics.

An EDC encoder creates a dependence (correlation) between successive digits of code words or sequences. The decoder makes use of this, when decoding, to detect and correct errors. The dependence also exists in the sequence of modulated signal elements, however, and this can be further exploited to improve the demodulation process. The best performance is achieved when the processes of demodulation/detection and decoding are carried out as one integrated operation. What is required is a code matched filter; the practical way to achieve this is to use *soft-decision detection* [6-9]. In effect the received demodulated baseband signal is quantised, and then decoded by means of one of the previously mentioned decoding algorithms but using a soft distance metric (Euclidean distance) instead of Hamming distance. In AWGN, a decoding gain advantage of about 2dB is obtained. A further advantage is achieved if the EDC code is matched to the modulation scheme, as in the case of correlative PSK and other combined coding-modulation schemes [11]. In this way, coding gains of up to 6dB are possible. It is of interest to note that implementation of the schemes does not require much additional complexity; in fact, certain schemes are simpler than schemes with equivalent performance which do not combine coding and modulation.

When assessing the relative performance of block and convolutional codes, it is appropriate to compare codes with similar block length and decoding search length (approximately 5hn), respectively. On this basis, there is not much difference in performance, see Fig. 4.2, and the choice will be made on other considerations. Some of the advantages of both types of code are achieved by using a *concatenated coding* scheme, involving an inner short constraint length convolutional code, with soft-decision decoding, and an outer RS code [6-9,12].

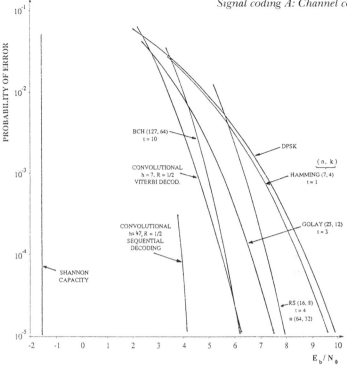

Fig. 4.2   Performance of (approximate) half rate codes
in additive white Gaussian noise (AWGU)

Errors occur in a diffuse mixture of burst and random patterns on mobile radio channels [3]. To enable the design of a really effective EDC scheme, what is required is some knowledge of the error pattern characteristics, and the existence of a set of suitable diffuse and burst error control codes. Even if some knowledge of the former could be obtained (from channel measurements or more refined channel models), the latter would prove to be a problem, since this type of EDC code is not well developed. The best current solution is therefore to use random EDC codes with a sufficient degree of interleaving; in the future, however, more efficient burst and diffuse codes may emerge.   Block codes are preferable to convolutional codes if messages are short because a 'tail' of hn bits is required at the end of the encoded message. Many practical error-control schemes use a combination of block and convolutional codes in a hybrid FEC-ARQ scheme; for example, the CDLC standard in the current UK TACS cellular system [3], and in the GSM scheme [13].   A recent survey of coded modulation schemes for fading channels [14,15] indicates the value of combining EDC coding and modulation in mobile radio systems.   It is also efficient to combine source and channel coding [16,17].   With careful EDC design, output (decoded) error rates in the range 10^{-3} to 10^{-7} are achievable with moderate system complexity.   As simpler decoding algorithms are devised, and VLSI technology develops, better performance will become feasible for future mobile systems.

Acknowledgement

Contributions to this section from Dr M. Beale and Mr D. J. Tait are gratefully acknowledged.

4.5   References on channel coding

1.   SHANNON, C. E., and WEAVER. W.: 'The mathematical theory of communication' (2nd Edn., Illini Books, 1963)

2.   KOBAYASHI, H.: 'A survey of coding schemes for transmission or recording of digital data', IEEE Trans. Comm. Tech., 1971, COM-19, (6), pp. 1087-1100

3.   MUNDAY, P.J.: 'Data communications over cellular radio', in BREWSTER, R.L. (Ed): 'Data communications and networks 2' (IEE Telecommunications Series 22, 1989)

4.   O'REILLY, J. J., and POPPLEWELL, A.: 'Class of disparity reducing transmission codes with embedded error protection', IEE Proc., 1990, Vol 137, Part I, No 2, April, pp 73-77

5.   BEKER, H., and PIPER, F.: 'Cypher systems - the protection of communications' (Northwood, 1982)

6.   LIN, S., and COSTELLO, D.J.: 'Error-control coding: fundamentals and applications' (Prentice-Hall, 1983)

7.   FARRELL, P.G.: 'A survey of error control codes' (in LONGO, E. (Ed) 'Algebraic Coding Theory and Applications', Springer-Verlag, 1979)

8.   MICHELSON, A. M., and LEVESQUE, A. H.: 'Error-control techniques for digital communication' (Wiley, 1985)

9.   CLARK, G. C., and CAIN, J. T.: 'Error-correction coding for digital communications' (Plenum, 1985)

10.   COMROE, R. A., and COSTELLO, D. J.: 'ARQ schemes for data transmission in mobile radio systems', IEEE Jour on Selec. Areas in Comms, 1984, Vol SAC-2, No 4, July, pp. 472-481

11.   UNGERBOECK, G.: 'Channel coding with multilevel/phase signals', IEEE Trans. Info. Theory, 1982, IT-28, (1), pp. 55-67

12.   VUCETI, B., and LIN, S.: 'Block coded modulation and concatenated coding schemes for error control on fading channels', Int. Conf. AAECC 7, June, 1989, Toulouse, France, pp. 26-30

13.   HODGES, M. R. L.: 'The GSM radio interface', Brit. Telecom Tech. Jour., 1990, Vol 8, No 1, January, pp. 31-43

14.   FARRELL, P. G.: 'Coded modulation for fading channels', Second Bangor Comms Symp., 1990, May, pp. 23-24

15.   SCHLEGEL, C., and COSTELLO, D. J.: 'Bandwidth-efficient coding for fading channels', IEEE Trans, SAC-7, 1989, December, pp. 1356-1368

16.   WYRWAS, R. R., and FARRELL, P. G.: 'Joint source-channel coding for raster document transmission over mobile radio', Proc IEE, 1989, Vol 136, Pt I, No 6, December, pp. 375-380

17.   HAGENAUR, J.: 'Rate-compatible punctured convolutional codes and their applications', IEEE Trans, 1988, Vol VOM-36, No 4, April, pp. 389-400

Signal coding B: Speech coding

Professor Costas Xydeas

4.6 Introduction

The performance of a speech codec (digitisation process) employed in a digital speech transmission/storage system determines, to a large extent, the recovered speech quality and capacity of the system. As a result of this and the rapid expansion in digital speech transmission, store-and-forward systems and services, efficient speech coding techniques have assumed considerable importance.

Speech coding has come a long way since the early days of pulse code modulation, i.e. direct quantization. Present day techniques seek to exploit the intrinsic properties of speech signals in order to remove redundancy and thus achieve improved compression/speech quality performance. This chapter is a brief introduction to speech coding algorithms in general, with emphasis being given to the encoding of narrowband-telephone signals for network communication quality speech. The most important coding techniques and achievements are then discussed in the light of existing international coding standards. Finally, future directions are briefly considered in terms of anticipated speech technology applications and the coding algorithms needed to support these applications.

4.7 Coding requirements

Speech coding algorithms are developed and optimised to satisfy a number of application specific requirements. Obviously, the quality of the recovered speech signal plays a major role in the design of a codec. The objective signal-to-noise ratio (SNR) [1] can assist in evaluating and comparing the performance of systems operating at relatively high bit rates (> 16 kbits/sec). However, when SNR measures are applied to intermediate and low bit rate codecs they often fail to correlate well with the subjective quality of the decoded speech signal. As a result various subjective tests are employed to quantify output speech quality, with the Mean Opinion Score (MOS) [1,2] scale (1 to 5) test being the most popular.

When applied to narrowband-telephone speech (300 Hz to 3.4 kHz) an MOS score of greater or equal to 4.0 implies high, "network" quality speech (often referred to as "toll" quality). An MOS value in the range of 3.5 to 4.0 corresponds to "communication" quality, and is characterised by some degradation. Communication quality speech is acceptable in certain telephone applications, such as mobile radio and voice mail. MOS values in the region of 2.5 to 3.5 imply "synthetic" quality speech. This is highly intelligible with reduced naturalness and limited speaker recognisability. Synthetic quality is found in low bit rate ($\leq$ 2.4 kbits/sec) secure voice transmission systems.

Although telephone bandwidth (less than 3.4 kHz) speech is primarily used in digital speech communication systems, speech quality is considerably enhanced by increasing the input bandwidth. "Broadcast" quality speech relates to a bandwidth of 50 to 7 kHz and offers a dramatic improvement in naturalness and intelligibility (unvoiced sounds), when compared to conventional telephone bandwidth speech. Voice applications such as ISDN teleconferencing and loudspeaker telephones are based on broadcast quality speech and require codecs that can produce an MOS score équal or greater than 4.

Decoded speech quality is closely linked to two other codec design factors, namely output bit rate and codec complexity. In general, as the required bit rate decreases, codecs become more sophisticated and consequently more complex, in an attempt to retain high decoded speech quality. Obviously, algorithm complexity has a direct impact on implementation cost and power consumption. Nevertheless speech coding quality deteriorates at lower bit rates and the challenge always exists to design efficient low bit rate, high speech quality, low complexity codecs. Fig. 4.3 highlights speech quality as a function of output bit rate for the three classes of speech coding schemes, i.e. waveform coders, vocoders and hybrid systems.

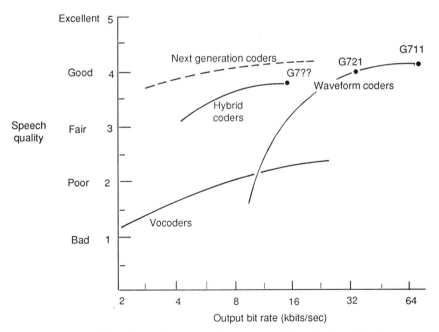

Fig. 4.3   Speech quality versus bit rate for telephone bandwidth speech codecs

In addition to quality, bit rate and complexity, a codec must satisfy other application specific requirements or characteristics in terms of codec channel error performance (robustness), codec delay, tandem coding effects and the encoding of voice-band data or other non-speech signals (signalling tones, music).

4.8 Coding techniques

Speech coding algorithms have been developed for encoding mainly telephone bandwidth speech and can be classified into three categories. The first class consists of algorithms which attempt to reproduce, at the output of the decoder, a close approximation of the original speech signal. They are known as *"waveform coders"* and operate successfully at intermediate and high bit rates (2 to 8 bits/ sample).

Pulse code modulation [1], the first waveform coding technique to be developed, consists of both time quantization (sampling) and amplitude quantization and although of limited compression/quality capability, PCM is used today at 64 kbits/sec, in both public and private "fixed" telecommunication networks, in the form of the G711 CCITT standard [3] with an MOS score of 4.3. The amplitude quantizer (or simply the quantizer) is an important element present in all encoding systems.

"Memoryless" and "adaptive" (uniform and non-uniform) scalar quantization [1] was studied during the sixties and seventies. Memoryless quantizers can be designed to provide maximum peak SNR for a given value of input power (optimum quantization) or an SNR performance that is almost independent of input power (logarithmic quantization). On the other hand, adaptive quantizers "track" the power of the input signal and provide an improved SNR performance within a large dynamic range. Adaptation can be "backward" or "forward", "block" or "sequential".

Forward block adaptive quantizers (AQF) [1, 3] are designed to "look ahead" in time, in order to derive their step size value which is subsequently used to quantize the block of incoming samples. As a result, AQFs introduce a certain processing delay, transmit periodically their step size as side information and exhibit a "robust" performance in the presence of channel errors.

Backward adaptive quantizers (AQB), employing block or sequential step size adaptation [1,4], derive their step size from "past" samples available at both the encoder and decoder. AQBs are therefore minimum delay schemes but can be sensitive to channel errors and thus require the introduction of "leakage" in their step size adaptation procedures.

An important generalisation of scalar quantization is Vector Quantization (VQ) [5, 6] where, instead of quantizing a single sample at a time, the quantizer now accepts "vectors" of samples. The input vector is usually gain normalised before it is compared to a set of vectors stored in a "codebook". The transmitted codeword, for a given input vector, is the index of that entry in the codebook which "best represents" the input vector.

Significant compression can be achieved only by allowing codecs to model (partially or fully) the short term characteristics of speech and by enabling the removal of redundancy from the signal prior to quantization. Redundancy removal methods decorrelate the input speech samples and reduce the energy of the signal to be quantized. This ensures that the codec produces lower quantization distortion, as compared to direct quantization. An improved performance, at a given bit rate, is therefore achieved or equivalently, the bit rate can be reduced without loss in speech quality.

Linear prediction (LP) is a valuable method for modelling both the short-term spectral envelope and the harmonic/pitch related spectral structure of speech. As such, LP has been used extensively in waveform coders, in general, and differential coding systems in particular, for the removal of "short" and "long" term correlations present in speech.

Differential coders form an error signal, as the difference between the input speech samples and corresponding prediction estimates, which is then quantized and transmitted. *Adaptive differential pulse code modulation* (ADPCM) [1, 7] and *adaptive predictive coding* (APC) [1, 8] represent two important intermediate bit rate (32 to 16 kbits/sec) differential codecs. Both systems estimate the incoming input samples using previously decoded samples.

ADPCM employs a short term predictor which models (partially) the speech spectral envelope. The predictor can be forward or backward adaptive with its coefficients defined periodically (block adaptive) or at every sampling instant (sequentially adaptive) [9].

The CCITT G721 ADPCM standard [10] achieves network quality speech (MOS score of 4.1) at 32 kbits/sec. This is a low complexity codec of reasonable robustness, when operating with channel bit error rates in the range of 10^{-3} to 10^{-2}, and is therefore well suited for wireless access applications based on low power handheld "communicators"[11].

G721 has been extended in recommendation G723 [10] to operate at 24 and 40 kbits/sec. In addition to G723 an "embedded ADPCM" standard has been established by CCITT [12] which operates at 40, 32, 24 and 16 kbits/sec and can be used in G.PVNP wideband packet network applications [13]. Notice that because speech quality deteriorates considerably at bit rates below 32 kbits/sec, "noise shaping" and "post-filtering" can be added to the codec in order to minimise the perceptual effect of quantization noise [14, 15].

APC employs both short and long term prediction in a differential coding structure. The system outperforms ADPCM at 16 kbits/sec and offers communication quality speech at bit rates as low as 10 kbits/sec. As with ADPCM, the introduction of noise shaping and postfiltering in APC reduces the subjective loudness of quantization noise. Finally improved APC performance can be obtained by using adaptive bit allocation [16], delayed decision coding [17] and vector quantization [18].

All the coding algorithms described so far, operate in the time domain. Redundancy removal techniques have also been applied successfully in the frequency domain (FD). FD waveform coding algorithms decompose the input speech signal into "sinusoidal" components, with time varying amplitudes and frequencies, and thus model speech as a time varying line spectrum. These are systems of moderate complexity and operate well at medium bit rates (16 kbits/ sec). When designed to operate in the range of 4.8 to 9.6 kbits/sec, the complexity of the approach used to model the speech spectrum increases considerably.

Medium to low bit rate FD waveform coding systems include:

Adaptive transform coding (ATC) [19, 20] coders derive a spectral representation of the short-term input signal and quantize the spectral coefficients using dynamic bit allocation. At the decoder, an inverse transformation accepts the received coefficients and generates the recovered speech signal. The dynamic bit allocation is usually based on an estimate of the spectral envelope of the signal and ensures that the signal to noise ratio is maximised. The bit allocation can also be adjusted to provide a "perceptually" optimised spectral distribution for the coding distortion present in the recovered signal and thus an improved coding performance. ATC offers near network quality at 16 kbits/sec. Output quality deteriorates rapidly, however, below 10 kbits/sec and low bit rate ATC systems (10 to 4.8 kbits/sec) employ adaptive bit allocation strategies based on spectral envelope and fine structure (pitch) information.

Sub-band coding (SBC) [21, 22] systems divide the input speech spectrum into a relatively small number of bands (2 to 16). This "wideband analysis" (in contrast to the narrowband block transformation ATC analysis) is performed using a "filterbank" approach. Each sub-band signal is then "translated" to zero frequency and sampled at its Nyquist rate. Again, as in ATC, by dynamically allocating a different number of bits for the encoding of each band, the variance of the reconstruction error can be minimized or alternatively the shape of the error spectrum can be "perceptually" optimized.

Sub-band coders produce near network quality speech at 16 kbits/sec, whereas at 9.6 kbits/sec speech quality is reduced as the "effective" bandwidth of the recovered signal is decreased. Communication quality speech can be obtained in the bit rate range of 4.8 to 8 kbits/sec by using sophisticated time domain coding for the sub-band signals and high frequency regeneration of "inactive" frequency bands [23].

SBC is recommended by CCITT for the coding of 0 to 7 kHz audio signals. In particular, CCITT standard G722 [24] is a 64 bits/sec SBC system developed primarily for coding 7 kHz input bandwidth speech in ISDN teleconferencing and loudspeaker telephone applications. The algorithm is basically a two band sub-band coder employing a fixed bit allocation strategy and ADPCM coding of the sub-band signals. G722 is a relatively low complexity coder (implementable on a single fixed point arithmetic DSP device) which offers low coding delay (3 msecs) and a robust performance with (i) random channel errors (bit error rate 10^{-4}) and (ii) tandeming (up to four encoding-decoding tandem stages). In addition to the 64 kbits/sec mode of operation, the algorithm can operate also at 56 and 48 kbits/sec with good speech quality (MOS score close to 4) [25].

Harmonic coding (HC) [26,27] algorithms perform a short-term Fourier analysis on the speech signal and "estimate" the harmonics of the line spectrum with the aid of a pitch predictor. An "estimated" spectrum (ESP) is reconstructed and subtracted from the original input spectrum. The resulting residual spectrum is encoded using ATC and transmitted together with parameters which define ESP. Harmonic coders deliver communication quality speech at 8 kbits/sec. Modifications to the basic HC model and dynamic quantization strategies have been proposed for operation at 6 and 4.8 kbits/sec [28].

Sinusoidal transform coding (STC) [29, 39] systems decompose the speech signal into time varying sinusoidal components (using short-term Fourier analysis) and employ a functional description of the time evolution of the amplitudes and phases of these components. Frequency tracks are reconstructed in each frame, preserving "continuity" between analysis frames and allowing the "death" of old and the "birth" of new frequency tracks. Cubic polynomials are used to provide "maximally smooth" phase unwrapping and frame boundary continuity. STC coders produce communications quality speech in the range of 8 to 4.8 kbits/sec.

An interesting extension of STC is the analysis-by-synthesis sinusoidal coder [31]. This scheme employs a polynomial representation of the time evolution of amplitude and phase components and determines the polynomial coefficients by minimizing the energy of the error signal formed as the difference between the input and decoded signals. The minimization process is appropriately constrained to produce "smooth" parameter tracks and signal continuity at the analysis frame boundaries.

The second class of speech coding techniques consists of algorithms called *"vocoders"* [32] which attempt to describe the speech production mechanism in terms of few independent parameters serving as the information-bearing signals.

Vocoders consider that speech is produced from a "source-filter" arrangement, see Fig. 4.4. Voiced speech is the result of exciting the vocal tract (filter) with a series of quasi-periodic glottal pulses generated by the vocal cords (source). Unvoiced speech, on the other hand, is produced by exciting the filter with random white noise. Thus, vocoders operate on the input signal, using an "analysis" process based on a particular speech production model, and extract a set of source-filter parameters which are encoded and transmitted. At the receiver, they are decoded and used to control a speech synthesizer which corresponds to the model used in the analysis process. Provided that all the perceptually significant parameters are extracted, the synthesised signal, as perceived by the human ear, resembles the original speech signal.

The "filter" part of the speech production model can be defined by operating in the time or frequency domain and effectively determines the "envelope" information of the speech spectrum. Typical frequency domain schemes are the *channel* [32, 33] and *formant* [32, 34] vocoders, whereas the popular *LPC vocoder* [32, 35] can be viewed as a time domain algorithm.

Vocoders are medium complexity systems and operate at low bit rates, typically 2.4 kbits/sec, with synthetic quality speech. Their poor quality speech is due to (i) the oversimplified "source" model used to drive the "filter" and (ii) the assumption that the source and the filter are linearly independent.

The inadequacy of the voiced-unvoiced (binary) vocoder excitation model is easily understood while attempting to classify short segments of speech into voiced and unvoiced segments. Although there are many easily recognised voiced/unvoiced segments, there are also cases where a clear decision is difficult and the excitation can be only characterised as "mixed". Furthermore it is known that apart from the main excitation pulse, which occurs at glottal closure, there is also "secondary" excitation, during a pitch period, as well as an interaction between excitation source and vocal tract.

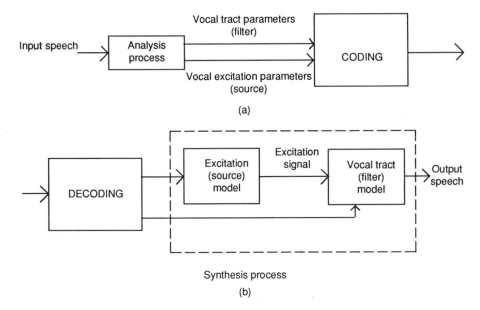

Fig. 4.4   Generalised block diagram of a vocoder
(a) Analysis at the transmitter
(b) Synthesis process at the receiver

In order to avoid the difficulties associated with devising an accurate excitation model, *hybrid coders* form their "excitation" signal directly from the input signal, usually by removing the spectral envelope information. The resulting "residual" signal is then coded, with the filter information (parameters), and transmitted to the receiver where speech is synthesised. Thus hybrid coders retain only the relatively accurate "filter" part of vocoders, which, when excited by an appropriately derived residual signal, can produce improved quality speech. Obviously, the bit rate and the output speech quality of a hybrid system depend on the methods used to define and encode the residual/excitation signal.

The "residual excitation" approach to speech coding was proposed originally with the *voice excited channel vocoder* (VECV) which is an improved speech quality, higher bit rate extension of the channel vocoder.

The system extracts a low frequency band from the input signal (250 to 940 Hz) which is waveform coded and transmitted, in addition to the vocoder channels (filter information). This "baseband" signal, which contains the required excitation information , is processed at the receiving end by a non-linear element that flattens and broadens the signal's spectrum, without affecting its periodicity (if any), to yield an "improved" excitation signal. However, the improved speech quality obtained from the voice excited channel vocoder (and from other residual excitation systems) is achieved only at the expense of several extra kbits/sec needed to code the baseband (residual) signal.

Many other residual excited hybrid coding systems have been proposed, most of them using linear prediction modelling of the synthesis filter. In particular, *residual excited linear predictive* (RELP) coding systems [36, 37] were studied extensively in the seventies for low to intermediate bit rate (4.8 to 16 kbits/sec) operation.

RELP systems employ short-term (and in certain cases, long-term) linear prediction, to formulate a difference signal (residual) in a "feed forward" manner. Early systems used "baseband" coding and transmitted a low pass version of the residual. The decoder recovered an approximation of the full band residual signal, by employing high frequency regeneration (HFR), which was subsequently used to synthesize output speech.

Simple HFR techniques, i.e. full-wave rectification, spectral folding and spectral translation, generate considerable distortion in the recovered speech signal. However, improved RELP systems, capable of producing communications quality speech at 8 kbits/sec have been proposed, which employ either pitch-aligned HFR procedures or full band pitch prediction in the time domain, to remove the pitch information from the residual signal prior to band-limitation/decimation [38,39].

The quality of the recovered speech signal can be improved considerably (at bit rates below 9.6 kbits/sec) if, instead of the "feedforward" RELP approach, an Analysis by Synthesis (AbS) optimisation procedure is used to define the excitation signal. This approach leads to the general *AbS predictive coder* of Fig. 4.5. In these systems both the "filter" and the "excitation" are defined on a short-term basis using a "closed-loop" optimisation process which minimises a "perceptually" weighted error measure formed between the input and the decoded speech signals.

AbS predictive systems can be classified into different groups according to the methodology used to define and code the excitation signal. These are:

Multipulse excitation linear prediction coders (MPE-LPC) [40]. MPE-LPC systems model the excitation signal as a sequence of q irregularly spaced pulses. That is, given a frame of n speech samples the n-samples excitation sequence contains q non-zero pulses located at sample instances $p_1, p_2, ... p_q$ with amplitudes $b_1, b_2, ... b_q$ respectively. MPE coders (as all the other AbS predictive schemes) employ a synthesis filter which consists of one or two autoregressive (AR) filters in series. The first filter models the "smooth" spectral envelope of the signal (short-term filter) while the second (if used) models the harmonic (fine) structure of the spectrum (long-term filter). The parameters of the excitation model (i.e. the pulse positions and amplitudes) and part of the synthesis filter (i.e. the long-term filter) are determined in a closed-loop optimisation process.

Various algorithms have been proposed for the optimisation of the excitation parameters and many of them are simple enough to permit real-time implementation [41,42].

When the pulse positions are fixed (known) the pulse amplitudes are determined by minimising the energy E of a "perceptually" weighted error formed between the input and synthesised speech signals. This involves the solution of a set of normal equations and imposes, effectively, an orthogonality condition between the synthesised speech and the error signals. The search to define the "optimum" (in a minimum error energy E sense) pulse positions $p_1, p_2, ... p_q$ can be a computationally demanding task since n >> q and the number $\left[\begin{smallmatrix}n\\q\end{smallmatrix}\right]$ of possible sets of positions is extremely large. Two broad classes of pulse positions estimation methods can be defined. These are successive elimination techniques (SE) and multivariate optimisation techniques (MO).

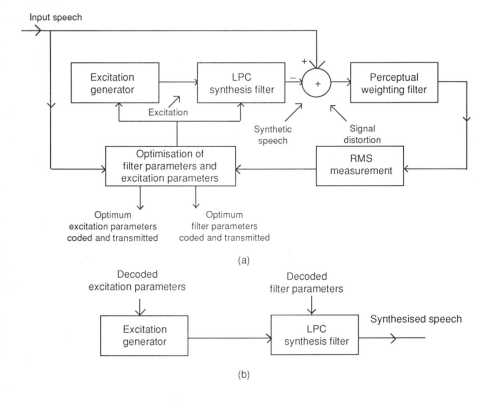

(a)

(b)

Fig. 4.5   Encoder (a) and Decoder (b) of a general
Analysis-by-Synthesis (AbS) predictive coder

Successive elimination techniques progressively decompose the "parameter space" of possible position sets into increasingly smaller subsets. The minimum of the error energy $E(p_1, p_2, ... p_q)$ is constrained by the subset boundaries and becomes more and more localised as the subsets shrink and finally reduce to single vectors. This is performed in a systematic way by restricting the range of values over which each of the position variables $p_1, p_2, ... p_q$ is allowed to vary. A repetitive evaluation of the approximation error function E is performed and decisions concerning the subdivisions of the subsets are taken according to the past history of E values. The complexity of SE techniques is determined by the number of iterations necessary to converge to a minimum error energy E_{min} solution and by the effort needed to compute E. SE methods include combinational search [43], branch and bound optimisation [44, 45] and multi-stage optimisation [46, 47, 48].

Multivariate optimisation techniques adopt a propabilistic approach for finding an optimum set of pulse positions. The "parameter space" ψ is irregularly "sampled" and information obtained from the "sampling" operation is used to direct the search towards the minimum of the error function E. Successive "samples" create paths in ψ which are progressively constrained, not in an absolute manner but in the sense that the probability of "sampling" a point (set of positions) outside a certain subset of ψ (which contains a minimum error energy point) becomes increasingly small. Examples of MO techniques are simulated annealing [49] and block search optimisation [50].

MPE-LPC coders provide near network quality speech (MOS≈4) at bit rates in the range of 16 to 8 kbits/sec. Their performance deteriorates rapidly however at bit rates below 8 kbits/sec where acceptable performance can only be achieved by drastically modifying the basic multipulse excitation model [51]. Notice that an MPE-LPC 9.6 kbits/sec codec has been recommended for use in the Skyphone airline application [52]. A special case of MPE-LPC coding is the regular-pulse excitation (RPE) LPC coder [53] which models the excitation signal with a sequence of equally spaced pulses. The performance of RPE systems is similar to that obtained from MPE coders. Various computationally efficient RPE schemes have been proposed and one of them has been chosen as the coding standard for the "full rate channel" of the GSM European mobile radio system [54]. The codec operates at 13.2 kbits/sec with a delay of 40 msec and a reasonably robust performance in the presence of channel errors. Both the 9.6 kbits/sec Skyphone and the 13.2 kbits/sec GSM algorithms are implementable on a single DSP device.

Codebook excitation linear prediction (CELP) coders [55] employ a vocal tract LP based model, a codebook based excitation model and an error criterion which serves to select an appropriate excitation sequence using an AbS optimisation process. Thus the system selects that excitation sequence which minimises a "perceptually" weighted mean square error formed between the input and the "locally" decoded signals. The vocal tract model utilises both a short term filter (STF), which models the spectral envelope of speech, and a long term filter (LTF), which accounts for pitch periodicity in voiced speech. The LTF can be defined "outside" or "inside" the AbS process by minimising (i) the energy of the "second residual" (i.e. the signal obtained by feeding speech through the inverse short term and long term filters) or (ii) the energy of the weighted error between the input and the decoded speech, respectively. In the latter case, the function of the LTF can be viewed as equivalent to an "adaptive excitation" codebook [56].

The main (fixed) excitation codebook was originally designed as a collection of random vectors (sequences) [55] each of which is constructed using samples from a set of independent identical distributed Gaussian random variables having zero mean and unit variance. In this case the number of computations required to select the "optimum" excitation sequence is prohibitively large and thus various simplified search strategies suitable for real time implementation have been proposed. These codebook search simplifications are based on (i) approximations in the way that the mean squared error is calculated [57] and (ii) modifications of the structure of the codebook. "Efficient" random codebooks with a "modelling" performance that is equivalent to that of the Gaussian random codebook include the space [58] and the ternary [59] codebooks. Considerable computational efficiency can be also achieved using structured codebooks [60, 61] and multistage codebook search strategies [62].

In addition to the above modifications whose aim is to improve computational efficiency, CELP speech quality can be enhanced by employing long term filters with high temporal resolution (HLTP) [63] and/or some form of post-filtering [56, 61] CELP coders provide very good (communication) quality speech at 8 kbits/sec, with a typical MOS score of 3.7 while at 4.8 kbits/sec CELP is far more successful than MPE-LPC in producing communication quality speech. The VSELP, CELP system, has been recently confirmed as the 8 kbits/sec North American standard for cellular telephony. The same algorithm has also been adopted at 6.7 kbits/sec [64] for the Japanese digital mobile radio system. The USA-DoD (Department of Defence) 4.8 kbits/sec speech coding standard [56] is also a CELP type system with speech quality comparable to 32 kbits/sec CVSD speech.

It is worth noticing that recent CCITT investigations into low-delay (≤ 5 ms) network-quality speech coding at 16 kbits/sec are focussed on a backward-adaptive CELP type coder with an MOS score of 4.0. The algorithm employs a 50 tap all pole synthesis filter the coefficients of which are updated every 6 m sec in a "backward adaptive" manner. The coder transmits a 7 bits codeword and a 3 bits gain every 5 samples and, although the "excitation search" process is quite complex, the algorithm is currently implementable in a full-duplex mode using two DSP devices.

Backward excitation recovery linear prediction coders (BER-LPC) [65, 66] employ a single or a multi-input speech synthesis filter and define the excitation signal(s) from past information which is already available at both the transmitter and the receiver. The parameters of the synthesis filter are determined by a close-loop optimisation procedure with minimises the energy of a "perceptually" weighted error formed between the original and the synthesised speech samples. Various "excitation adaptation" and filter optimisation algorithms have been proposed and certain algorithms lead to codecs with very small encoding delays (≈ 3 msecs). BER coders operating in the region of 8 to 4.8 kbits/sec offer similar speech quality to that obtained from CELP systems.

Finally, the MPE, CELP and BER excitation models can be combined to form hybrid AbS systems [67,68]. Fig. 4.6 shows the various speech coding methods.

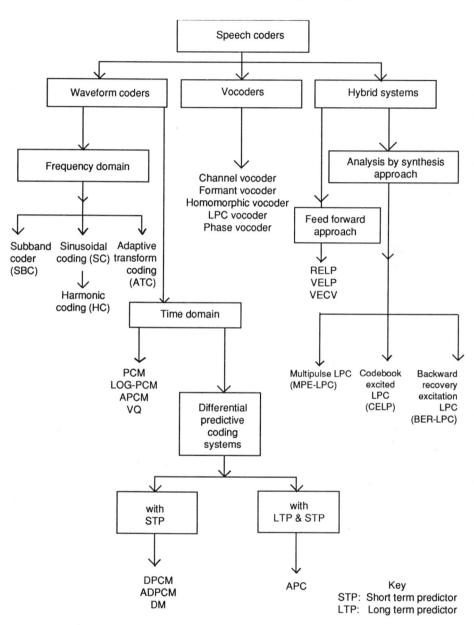

Fig. 4.6   Speech coding methodologies

4.9 Speech coding applications and directions

Medium and low bit rate hybrid coders are viewed as key "enabling" elements in the introduction of new voice services/applications. At 16 kbits/sec the CCITT list of possible applications [69] is quite long with entries such as: video services, cordless telephones, digital satellite systems, ISDN, packetised speech, store-and-forward systems, and voice messages for recorded announcements. Algorithmic speech coding technology is already in a position to offer solutions to these applications. Future work at this bit rate will be concerned mainly with finding trade-offs between coder quality performance and complexity, so that algorithms satisfying certain application specific requirements could be also implemented economically with existing DSP devices.

The major applications for 8 to 4 kbits/sec coders are mobile telephony/personal communications, mobile satellite communications (MSAT, INMARSAT etc.) and secure government and military communications. Also, 8 to 4 kbits/sec coders are extremely useful in private communication networks employing digital lines where many voice channels can be packed within 64 kbits/sec. As network quality is an important requirement in many of the above applications, particularly in mobile telephony and other systems which interface to the fixed public network, speech coding research is currently focussed on the development of 4.0 to 8.0 kbits/sec network quality algorithms.

The development of the next generation speech codecs, see Fig. 4.3, is linked to a better understanding of speech perception and noise masking, coupled with progress in efficient analysis by synthesis optimisation algorithms. These algorithms will be complex but future advances in DSP device technology will be able to support the ever increasing complexity of low bit rate codecs.

4.10 References for speech coding

1   JAYANT, N.S., and NOLL, P.: 'Digital coding of waveforms', (Prentice Hall, 1984)

2   KITAWAKI, N. and NAGOBUCHI, H.: 'Quality assessment of speech coding and speech synthesis systems', IEEE Comms Mag., 1988, Oct, pp. 36-44

3   EVCI, C. C., XYDEAS, C. S., and STEELE, R.: 'DPCM with adaptive forward quantization using second order adaptive predictors for speech encoding', IEEE Trans. on ASSP, 1981, Vol. ASSP-29, No. 3, June

4   XYDEAS, C. S., STEELE, R., and FARUGUI, M. N.: 'Envelope dynamic ratio quantizer', IEEE Trans. on Comms, 1980, Vol. Com-28, No. 5, May

5   MAKHOUL, J., ROUCOS, S., and GISH, H.: 'Vector quantization in speech coding', Proc IEEE, 1985, Vol. 73, No. 11, Nov

6 SABIN, M. J., and GRAY, R. M.: 'Product code vector quantizers for waveform and voice coding', IEEE Trans. on ASSP, 1984, Vol. ASSP-32, No. 3, June

7 XYDEAS, C. S., EVCI, C. C., and STEELE, R.: 'Sequential adaptive predictors for ADPCM speech encoders', IEEE Trans. on Comms, 1982, COM-30 No. 8, Aug

8 ATAL, B. S.: 'Predictive coding of speech at low bit rates', IEEE Trans. on Comms, 1982, COM-30 No. 4, April

9 XYDEAS, C. S., and EVCI, C. C.: 'A comparative study of DPCM-AQF speech coders at 16 to 32 kbits/sec', IEEE Inter. Conf. on ASSP, 1982, May

10 CCITT Study Group XVIII, '32 kbits/sec ADPCM', Working Party 8, Draft revision of Recommendation G721, Draft Document No. D 723/XVIII, Geneva, Switzerland

11 STEELE, R.: 'The cellular environment of lightweight handheld portables', IEEE Comms Mag. 1989, July, pp 20-29

12 CCITT, 'Draft Recommendation G. EMB - 5, 4, 3 bit per sample embedded ADPCM' Annex 1, Appendix 1 to Question (24/XV)

13 CCITT, 'Draft Recommendation G. PVNP - packet voice networking protocol', Annex 1, Appendix 2 to Question (24/XV)

14 XYDEAS, C. S. and YEOH, F. S.: 'Noise spectral shaping applied to course quantization differential speech coders', IEEE Proc. Melecon-83, 1983, Vol. 11, May

15 RAMAMOORTHY, V., JAYANT, N. S., COX, R. V., and SONAHI, M. M.: 'Enhancement of ADPCM speech coding with backward-adaptive algorithm for postfiltering and noise feedback', IEEE J. Sel. Areas in Comms, 1988, Feb, pp 364-382

16 HONDA, M., and HEKURA, F.: 'Bit allocation in time and frequency domains for predictive coding', IEEE Trans. on ASSP, Vol. 32, No 3, 1989, June, pp 465-473

17 IYENGER, V., and KEBAL, P.: 'A low delay 16 kbits/sec speech coder', IEEE Proc. ICASSP, 1988

18 CHEN, J., and GERSHO, A.: 'Real time vector APC speech coding at 4.8 kbits/sec', IEEE Proc. ICASSP, 1987

19 ZELINSKI, R., and NOLL, P.: 'Adaptive transform coding of speech signals', IEEE Trans. on ASSP, 1977, Vol. 25, No 4, Aug

20   CROCHIERE, R. et al: 'Real time speech coding', IEEE Trans. on Comms, 1982, April, pp. 621-634

21   YEOH, F. S., and XYDEAS, C. S.: 'Transform approach to split band coding schemes', IEE Proc. 1984, Feb, Vol. 131, Part F No 1

22   COX, R., et al: 'New directions in sub-band coding', IEEE J. Select Areas in Comms, 1988, Vol. 6, No 2, Feb

23   SAVVIDES, V., and XYDEAS, C. S.: 'A new approach to low bit rate speech coding', Proc. Int. Conf. on Digital Processing of Signals in Comms, IERE, Loughborough, U.K. Sept. 1988

24   CCITT Study Group XVIII, '7 kHz audio coding within 64 kbits/sec', CCITT Draft Recommendation G72x, Report of Working Party XVIII 8, July 1986

25   MODENA, G., COLEMAN, A., USAI, P., and COVERDALE, P.: 'Subjective performance evaluation of the 7 kHz audio coder', IEEE, Proc. Globecom, December 1986

26   THOMSON, D. L.: 'Parametric models of the magnitude/phase spectrum for harmonic speech coding', IEEE Int. Conf. on ASSP, 1988

27   TRANCOSO, M. et al: 'Quantization issues in harmonic coders', IEEE Intern. Conf. on ASSP, 1988

28   BRONSON, E. C. et al: 'Harmonic coding of speech at 4.8 kbits/sec', IEEE Intern. Conf. on ASSP, 1987

29   McAULAY, R. J., and QUATIERI, T. F.: 'Speech analysis/ synthesis based on sinusoidal representation', IEEE Trans. on ASSP, August 1986, pp 744-754

30   McAULAY, R. J., and QUATIERI, T. F.: 'Multirate sinusoidal transform coding at rates from 2.4 to 8 kbits/sec', IEEE, Intern. Conf. on ASSP, 1987

31   GEORGE, E. B. and SMITH, M. J. T.: 'A new speech coding model based on least squares sinusoidal representation', IEEE, Intern. Conf. on ASSP, 1987

32   FLANAGAN, J. L.: 'Speech analysis, synthesis and perception', (Springer-Verlag, 1965)

33   DUDLEY, H.: 'The vocoder', Bell Laboratories Record 17, 1939, pp. 122-126

34   COKER, G. H.: 'Computer simulated analyser for a formant vocoder', JASA, Vol 35, 1963

35 MARKEL, D. J., and GRAY, A. H.: 'Linear prediction of speech', (Springer-Verlag, New York, 1976)

36 VISWANATHAN, R., et al: 'Voice excited LPC coders for 9.6 kbits/sec speech transmission', IEEE Intern. Conf. on ASSP, 1979

37 UN, C. K., and LEE, J. K.: 'On spectral flattening techniques in RELP', IEEE Intern. Conf. on ASSP, 1982

38 KATTERFELDT, H.: 'A DFT-based RELP for 4.8 and 9.6 kbits/sec', IEEE Intern. Conf. on ASSP, 1981

39 SLUYTER, R. et al: 'An efficient pitch-aligned high-frequency regeneration technique for RELP vocoders', IEEE Inter. Conf. on Comms., May 1984

40 ATAL R. S., and REMDE, J. R.: 'A new model of LPC excitation for producing natural sounding speech at low bit rates', IEEE Proc. ICASSP, 1982

41 GOUVIANAKIS, N., and XYDEAS, C.: 'A comparative study of multistage sequential and block sequential search multipulse LPC algorithms', Proc. Int. Conf. IASTED, Paris, June 1985

42 KUKUI, A., and SHIBAGAKI, K.: 'Implementation of a multipulse speech codec with pitch prediction on a single chip floating-point signal processor', IEEE Proc. ICASSP, 1987

43 DEMBO, A., and MALAH, D.: 'A new approach to multipulse LPC coder design', IEEE, Proc. ICASSP, 1985

44 LANG, S. W.: 'Solving a class of nonlinear least squares problems', IEEE Proc. ICASSP, 1985

45 FOULDS, L. R.: 'Optimisation techniques, an introduction', (Springer Verlag, New York, 1981)

46 BEROUTI, M. et al: 'Efficient computation and encoding of multipulse LPC', IEEE, Proc. ICASSP, 1984

47 ARASEKI, T. et al: 'Multipulse excited speech coder based on maximum crosscorrelation aearch algorithm', IEEE Proc. Global Telecoms Conf., 1983

48 SINGHAL, S.: 'Reducing computation in optimal amplitude multipulse coders', IEEE, Proc. ICASSP, 1986

49 THORPE, T. F., and KINGSBURY, N. G.: 'A monotonic descent algorithm to choose pulse locations in the excitation sequence of a multipulse speech coder', IEE Electronic Letters, Vol. 21, No. 21, Oct. 1985, pp 972-973

50 GOUVIANAKIS, N., and XYDEAS, C.: 'A multipulse excited LPC coder implementation based on a block solution approach', IERE Proc. Int. Conf. on Digital Processing of Signals in Communications, Loughborough 1985, and U.K. Patent No. 8515501

51 ONO, S., and OZAWA, K.: '2.4 kbits/sec pitch prediction multipulse speech coding', IEEE Proc. ICASSP, 1988

52 BOYD, I., and SOUTHCOTT, C. B.: 'A speech codec for the Skyphone service', British Telecom Tech. Journal, Vol 6, No. 2, 1988

53 KROON, P. et al: 'Regular pulse excitation: a novel approach to effective and efficient multipulse coding of speech', IEEE Trans. on ASSP, Oct 1986

54 VARY, P. et al: 'Speech codec for the European mobile radio system', IEEE Inter. Conf. on ASSP, April 1988

55 SCHROEDER, M. R., and ATAL, B. S.: 'Code-excited linear prediction: high quality speech at low bit rates', IEEE Proc. ICASSP, 1985

56 CAMPBELL, J. P. et al: 'An expandable-error protected 4800 b/s CELP coder', (US Federal Standard), IEEE Inter. Conf. on ASSP, 1989

57 TRANCOSO, I. M. and ATAL, B. S.: 'Efficient procedures for finding the optimum innovation sequence in stochastic coders', IEEE Intern. Conf. on ASSP, 1986

58 DAVIDSON, G., and GERSHO, A.: 'Complexity reduction methods for vector excitation coding', IEEE, Intern. Conf. on ASSP, 1986

59 XYDEAS, C. S., IRETON, C. and BAGHBADRANI, D. K.: 'Theory and real time implementation of a 4.8/6 kbits/sec CELP coder using ternary code excitation', IERE, Proc. of 5th Inter. Conf. on Digital Processing of Signals in Comms., Loughborough, UK, September 1988

60 IRETON, M. A., and XYDEAS, C. S.: 'On improving vector excitation coders through the use of spherical lattice codebooks', IEEE Intern. Conf. on ASSP, 1989

61 GERSON, I., and JASIUK, M. A.: 'Vector sum excited linear prediction (VSELP) speech coding at 8 kbits/sec, IEEE Intern. Conf. on ASSP, 1990

62 BAGHBADRANI, D. K., XYDEAS, C. S., and MORLEY, S.: 'Single DSP high quality speech CELP at 8 to 4.8 kbits/sec', Signal Processing V: Theories and Applications, (Elsevier Science Publishers, B.V., 1990)

63 KROON, P., and ATAL, B. S.: 'Pitch predictors with high temporal resolution', IEEE Intern. Conf. on ASSP, 1990.

64 GERSON, I. A., TASIUK, M. A. et al: 'Combined speech and channel coding at 11.2 kbits/sec', Signal Processing V: Theories and Applications, (Elsevier Science Publishers B.V., 1990)

65 GOUVIANAKIS, N., and XYDEAS, C. S.: 'Advances in analysis by synthesis LPC speech coders', IERE Journal, Suppl. on Mobile Radio, No. 6, Nov/Dec. 1987 and UK Patent No. 8720388

66 ROSE, R. C., and BARNWELL, T. P.: 'The self-excited vocoder - an alternative approach to toll quality at 4.8 kbits/sec', IEEE Intern. Conf. on ASSP, 1986

67 ROSE, R. C., and BARNWELL, T. P.: 'Quality comparison of low complexity 4800 bits/sec. Self excited and code excited vocoders', IEEE Proc. ICASSP, 1987

68 SREENIVAS, T. V.: 'Modelling LPC-residue by components for good quality speech coding', IEEE Proc. ICASSP, 1988

69 CCITT, 'Possible applications for 16 kbits/sec voice coding', Appendix 3 - Annex 1 to Question 21/XV, 13-22, March 1989

[Ed. note: A further review of source coding of speech in connection with satellite services appeared in the July 1990 issue of the Proceedings of the IEEE, namely: SINGHAL, S., LE GALL, D., and CHEN, C-T.: 'Source coding of speech and video signals', Vol. 78, p.1233]

Chapter 5

Conventional private mobile radio

David A Hanson

5.1 Introduction

Private Mobile Radio (PMR) in the UK is a particular part of the Land Mobile Radio service. PMR users are businesses and organisations licensed through the Radiocommunications Agency (RA), an Executive Agency of the Department of Trade and Industry to set up and operate their own self-contained (hence 'private') mobile radio communications systems.

Typical users of PMR range from the smallest taxi company with an operating radius of a few kilometres, to national operators of vehicle fleets such as the roadside breakdown organisations. District and county councils, area health authorities (including the ambulance services), and the water supply industry are typical of systems where county-wide areas of coverage are required. In contrast construction sites, factories and shopping precincts are typical of locations where short range, usually handportable equipment would be used.

The Fuel and Power industries - Gas, Coal and Electricity - and some Emergency Services whilst having communications systems and equipment similar to PMR, occupy their own separately administered part of the spectrum. Police forces also have their own frequency allocations and additionally for operational reasons have both facilities and equipment not available in PMR.

5.2 Frequencies used in PMR

Fig. 5.1 shows how the spectrum at VHF and UHF is allocated to PMR [1] (see Annex A for more detail). Either amplitude modulation (AM) or frequency modulation (FM) is permitted at VHF although the world-wide trend is towards FM and consequently the limited availability and higher cost of AM equipment means that most if not all new systems operate with FM. Only FM may be used on UHF channels. In mid-band AM has predominated but is giving way to FM.

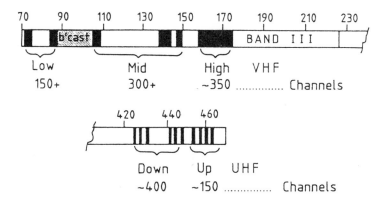

Fig. 5.1   Spectrum allocated to PMR (1990)

The licence issued to a PMR user by the RA specifies the frequencies of the channel or channels the user is to occupy [2]. Where these are in the spectrum depends on several factors including:

- historical: original business users of PMR tended to be in low band with mid-band exclusively allocated to fuel and power and emergency services.

- propagation: low band VHF is preferred for best coverage in rural areas such as by the water supply authorities. UHF is more efficient in the major urban areas and for shorter range systems UHF channels may be preferred (some cities have additional allocations down-band to meet demand), e.g. see previous Table 1.2.

- spectrum overcrowding: although prospective users may to some extent choose the band in which they operate, some may be directed to a part of the spectrum where channels are less congested. The RA are no longer allocating high band VHF channels in the south east of England [3].

The demand for PMR licences has increased considerably in recent years, and with more than 20,000 licencees, not including emergency services and other midband users, operating a total of over 460,000 mobiles (1988/89 figures [4]), the average business user will be expected to share the channel allocated with other companies in the same area [5].

5.3   Channelling in PMR

All PMR channels in the UK are separated at intervals of 12.5kHz. In determining the width of the transmission occupying each channel allowance has to be made for frequency drift (licenses quote up to ±1kHz for fixed units, ±2.5kHz for handportables) and for the non-ideal characteristics of the transmitting and receiving equipment. A maximum of 7kHz - 8kHz [6] is therefore available for occupation by a transmission, leaving around 5kHz between adjacent transmissions to allow for drift and include a guard band. The occupied bandwidth corresponds to a maximum modulating frequency of around 3kHz and, in systems, peak deviation of ±2.5kHz.

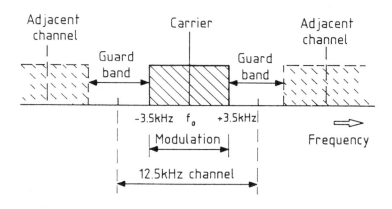

Fig. 5.2   PMR channelling arrangement

With 12.5kHz channels, the spectrum allocated to PMR (excluding Band III) therefore provides around 1000 channels for the whole of the UK.

5.4   PMR system

In a typical PMR system, basically indicated in Fig. 5.3, a fixed base station communicates with a number of mobile units which may be vehicle mounted or handportable; base station location, maximum transmitter power and number of mobile units being specified by the user's license. Where the system operates over a small area, direct contact between mobiles in addition to the base station - if there is one - is common.

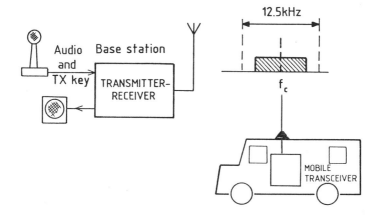

Fig. 5.3   PMR operating system

Before looking in more detail at the components of a typical system it should be mentioned that many categories of radio communications equipment, including all those used in PMR have to carry an approval number (the RTD number) issued by the RA. Specifications for type approval, codes of practice and a number of other aspects of PMR are published under a variety of MPT numbers [6 - 10, 17, 19, 23, 27] and manufacturers must submit sample equipment for approval testing against a specific MPT [11] in order that an RTD number be issued before the equipment is marketed.

5.4.1 A typical PMR transceiver

The transmitter:

The function of the transmitter is to supply a modulated RF carrier at the required frequency and power level to the aerial system, with spurious components including harmonics below the required levels [12]. The carrier frequency at either VHF or UHF needs to be precisely controlled to prevent drift and consequent loss of intelligence at the receiving end of the channel, or interference to the adjacent channels.

Conventionally, as shown in Fig. 5.4, the first stage is a quartz crystal oscillator which provides a stable and accurate output at a known frequency which is multiplied up to produce the carrier frequency and then amplified to the required power level before transmission via the aerial. At some point in this process modulation has to be introduced, the type of modulation determining where in the transmitter this takes place. In an FM system, the crystal frequency may be modulated directly using a varactor diode in the oscillator circuit or phase modulation applied in a following stage. In either case the deviation produced will take into account the frequency multiplication following and the final deviation required ($\pm$2.5kHz for 12.5kHz channelling).

With AM, the modulation is applied at the final carrier frequency and unless linear amplifiers were to be used (unlikely in PMR) the modulation process has to take place at the final RF power amplifier.

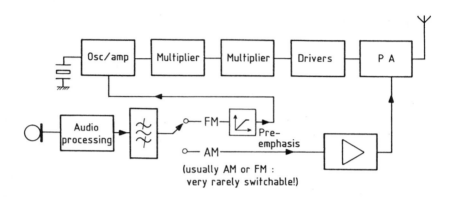

Fig. 5.4   A PMR transmitter arrangement

Audio processing in both AM and FM transmitters includes circuitry to ensure that a level, bandlimited signal (approx. 300Hz to 2.8kHz) signal is applied to the modulator. FM transmitter audio processing also includes pre-emphasis of higher audio frequencies for reasons connected with FM demodulator characteristics (see below).

The receiver:

The range of a communications system for a given transmitter power is limited by the ability of the receiver to reproduce an intelligible version of the original modulating signal. Modern PMR receivers are required to detect signals of 0.2uV - 0.3uV from the antenna (-120dBm to -117dBm) and produce an output of perhaps 1 watt of audio. The receiver must be uniformly selective wherever tuned in its specified range of operation - for example 160MHz to 174MHz at VHF or 440MHz to 465MHz at UHF. The receiver must also reject strong signals outside the wanted channel yet maintain sensitivity in the presence of these unwanted signals [13].

The superheterodyne (or superhet) receiver achieves the overall gain and selectivity required across a range of incoming frequencies by converting the required input signal frequency (RF) to a fixed 'intermediate' frequency (IF). At this fixed frequency, amplifiers and filters in the IF stages provide the high gain and precise bandwidth required. The conversion of incoming RF to IF is achieved by mixing the RF with a local oscillator (LO) signal generated in the receiver, whose frequency therefore determines which of the many incoming signals present at the aerial is accepted by the receiver. Conventionally the LO signal is provided by a crystal oscillator/frequency multiplication process, similar to the generation of carrier frequency in the transmitter and with the same degree of precision.

The filtering characteristics of the IF stages are critical to receiver performance and are best at a low IF where a steep-sided response with the required 7-8kHz passband is possible using ceramic filters. However the mixing process preceding the IF stages will produce an output at the IF from incoming signals at both the wanted RF and any signal present at the same interval on the other side of the IF:

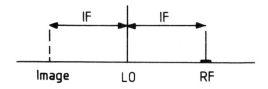

Fig. 5.5   The IF image

Any signal on the unwanted frequency, termed the IF 'image', has to be rejected prior to the mixer, usually by a tuned RF amplifier. For frequencies at VHF and UHF this requires a high IF and 10.7MHz or 21.4MHz are typical. The required IF filter response however is not easily achieved at these high values of IF, so that following a degree of amplification and filtering in the first IF stages, the signal is mixed again with a fixed frequency, second local oscillator down to another low IF, usually 455kHz, where the filter response is nearer to ideal and whose image is easily filtered out in the first IF stages, indicated in Fig. 5.6.

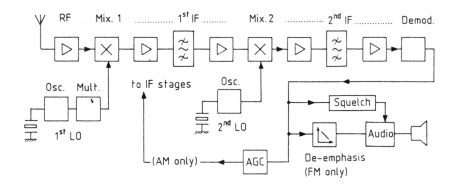

Fig. 5.6   A typical PMR double superhet receiver structure

The RF amplifier, as the first stage in the receiver, is also critical in determining the noise performance of the receiver as a whole. Dual gate FET's have made it possible for PMR mobiles to detect signals below -120dBm (0.2uV p.d.) at the antenna input.

AM receivers differ from FM receivers in that over the considerable range of input signal amplitudes possible AM systems must preserve amplitude variations through all stages of the receiver without overloading. AM receivers therefore have automatic gain control (AGC) applied to the IF stages (and, possibly, RF stage) which reduces their gain as the incoming signal level at the demodulator rises. In an FM receiver however the amplitude of all incoming signals presented to the demodulator should ideally be the same, so that hard limiting in the IF stages is actually required rather than avoided.

One additional requirement in FM systems follows from the rising noise vs frequency characteristic at the output of the FM demodulator. Applying pre-emphasis to higher modulating frequencies at the transmitter and corresponding de-emphasis to receiver audio counteracts this effect and a level S/N ratio is then achieved across the range of received audio frequencies.

PMR receivers also require squelch circuitry to cut off receiver noise at the audio output in the absence of an incoming transmission, so that a receiver on standby is silent until a signal of reasonable strength arrives.

Transmit-receive switching:
Though some base stations use a separate antenna for transmit and receive, single antenna working is universal with mobiles. T/R switching via a relay is now rare; solid state switching using PIN diodes is preferred, as shown in Fig. 5.7.

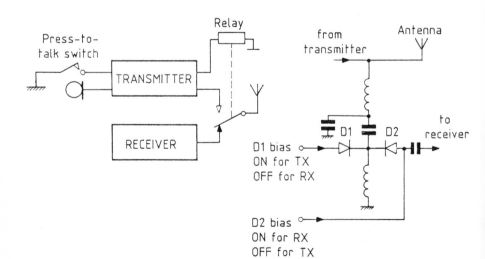

Fig. 5.7 Transmit/receive switching arrangements

5.4.2 Frequency synthesisers

A conventional single channel transceiver requires two crystals - one each to define the transmit and receive frequencies: these crystals have to be obtained from specialist suppliers to order for the user's channel. This arrangement is economical in the many single channel systems in use but where PMR equipment is required to operate through several channels, for example when roaming over a wide area covered by several base stations or to have frequent changes of channel when hired out to different users, a frequency synthesiser providing both carrier and first local oscillator frequencies is preferred which may be programmed to any required channel or channels.

5.4.3 Synthesiser operation

The output of a synthesiser is provided by a voltage controlled oscillator (VCO) Fig. 5.8 whose frequency is held at the required value by a control loop. The VCO output frequency is first divided down by a programmable divider, to the same nominal value as a fixed reference frequency. The divided VCO and reference frequencies are compared in a phase sensitive detector (PSD) whose output, when filtered - essentially a DC voltage representing any error between the two - is the control voltage of the VCO.

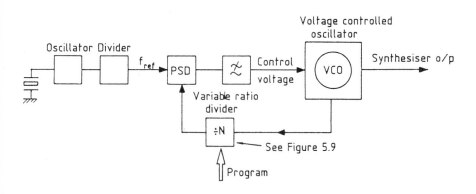

Fig. 5.8 The basic frequency synthesizer

If the programmable divider ratio is N, the VCO frequency when locked is N times the reference frequency so that changing N in steps of 1 will increment the VCO output in steps equal to the reference frequency. For PMR, the minimum increment in frequency required is the same as the channel spacing, i.e. 12.5kHz. This reference frequency would normally be provided by a crystal oscillator divided down: for example 12.8MHz divided by 1024, although a submultiple such as 6.25kHz is also found in some systems.

The ratio N may take values over a wide range: as an example, if the synthesiser above were capable of producing the transmitter frequency on any channel in VHF high band, say 160MHz - 174MHz, N would have to take any integer value between 12800 and 13920 with of course the divider input operating at these frequencies. One possible solution is to first divide the VCO output by a 'prescaler' using fast ECL followed by further dividers using CMOS. However if this prescaler had only one division ratio or modulus then the overall division ratio N could not take values in steps of 1. The rather neat solution is to use a dual modulus prescaler, with the two division ratios available differing by 1 (for example 41 and 40) the modulus in use being selected by two subsequent programmable counters, Fig. 5.9.

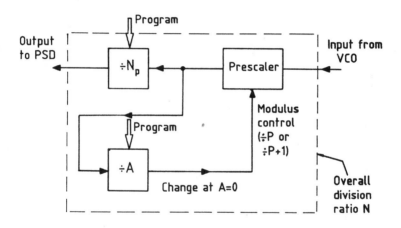

Fig. 5.9 A programmable divider with dual modulus prescaling

Both counters are loaded with binary numbers appropriate to the required final value of N (provided by the synthesiser programming using a diode matrix or PROM/EAROM), and the prescaler set to divide by P+1. Both the Np and A counters then count down until the A counter holds zero, on input pulses at the rate of F(VCO)/(P+1). At this point the prescaler modulus is reduced to P and the Np counter continues to count down to zero on pulses at the rate of F(VCO)/P. At the count zero both counters are reloaded, the prescaler reset and the countdown is repeated. The overall count N is therefore:

A(P+1) from the first stage + (N-A)P from the second stage.

or $N = P.N_p + A$

The size of the N_p and A counters therefore depends on the value of N. For example, a synthesiser arrangement with a 64/65 prescaler [14], requires that the N_p and A counters are 10 bit and 6 bit respectively in order to provide the range of values of N (maximum 40,000).

In synthesised FM equipment modulation is applied at the VCO, the VCO loop when locked having a response time which is sufficiently long to prevent the loop trying to follow the modulation. In AM equipment the modulation is applied at the final RF power amplifier.

5.4.4 Antenna Systems

The majority of VHF and UHF antenna systems [15] in PMR use are required to operate with vertical polarisation and to have omni-directional characteristics so that dipoles are the most common. Vehicles are usually fitted with either 1/4 or 5/8 wavelength elements, the latter with inductive loading for matching purposes, relying on the bodywork around the mounting point to provide a ground plane.

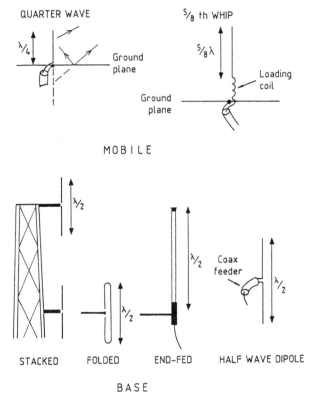

Fig. 5.10   Base and mobile antenna

Base station antenna are usually mounted on masts or towers and range from a centre or end-fed dipole to the most complex consisting of stacks or arrays of dipoles to increase gain in the vertical or horizontal plane as necessary for the system coverage required.

5.5   Modes of operation

- Single frequency simplex

(base-to-base co-channel
interference shown)

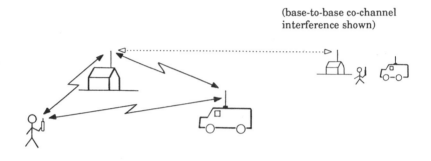

Fig. 5.11   Single frequency simplex operation

Fig. 5.11 illustrates an arrangement with a base station and mobile units. In simplex, traffic is one-way at any time, with a handover at the end of each period of transmission. In the single frequency situation all units hear all transmissions within range, which many users would find desirable.

The problem in this mode however is that of frequency re-use bearing in mind the demand for the limited number of channels available. If good coverage is required, the base equipment would be placed on an elevated site; another user of the same channel, perhaps even some distance away but also with a base station on a high site would hear all transmissions from the first base and vice versa. Since base stations are likely to make the majority of transmissions in most systems, considerable interference to incoming traffic from mobiles would occur due to simultaneous transmissions from other base stations. Mobiles of course would only receive transmissions from both bases in the region of overlapping coverage.

To reduce coverage area therefore and maximise re-use of channels, single frequency simplex is restricted to a transmitter effective radiated power (ERP) of 5 watts, and is mainly used with groups of handportable equipment on construction sites and around factories for example, and for short-term hire of handportable and vehicle-mounted equipment.

- Two frequency simplex

A higher degree of reuse without undue interference is achieved by using two frequencies for each channel, i.e. Fig. 5.12.

The two base stations are now unable to hear each other and overall interference is greatly reduced. A possible disadvantage is that mobiles are now unable to hear each other and this may often lead to two mobiles calling in at the same time in a busy system, although in larger systems it is possible to obtain approval for pip tones to be transmitted by the base while a mobile transmission is being received, indicating to all other mobiles that the channel is busy.

The separation between base and mobile transmit frequencies differs in the various parts of the spectrum (see Annex A). The degree of separation however is always such that at sites with several base stations, intermodulation products at the output of any one transmitter due to leakage from another are well away from base receive frequencies (Chapter 12).

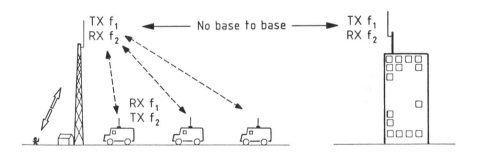

Fig. 5.12   Two frequency simplex arrangement

The arrangement in Fig 5.12 is the most common mode of operation in PMR; two frequency simplex with a dispatcher working through a base station controlling a group of mobiles. The maximum ERP in this case is usually 25 watts from both mobile and base stations.

- Talk-through

In a two frequency simplex channel the separation of base and mobile transmitter frequencies makes possible an arrangement where the base transmitter could re-transmit incoming received signals, with the receiver and transmitter isolated by notch filters. This mode known as 'talk-through' would enable mobiles to communicate directly. The problem created by the use of a talk-through facility of this type is that since channels are likely to be shared by more than one licensee, all mobiles regardless of affiliation calling within the range of such a talk-through base would be re-transmitted by that base creating unnecessary interference on the channel. In effect the channel suffers from the disadvantage of the single frequency channel, in terms of re-use and interference and for this reason a simple 'free-running' talk-through facility is not permitted in PMR.

5.6   Signalling systems in PMR

Overcrowding of the spectrum requires that most users share channels with others in the same area. Apart from sharing by several licensees, there are situations where different user groups in the same organisation may have to share the single channel allocated to them - in a large factory for example. Signalling techniques have therefore been devised [16] to isolate groups by coding their transmissions in some way so that each group does not hear the traffic of any other on the channel, although this does not of course prevent interference if two transmissions are received simultaneously.

Forms of signalling also exist which enable specific mobiles to be addressed so that they only hear transmissions intended specifically for them, and others which enable simple prearranged messages or instructions to be sent. The basis of such signalling systems in PMR is the use of audio tones which lie within the passband of the transmitter and receiver and are readily interfaced with the audio circuitry. The generation and control of such signalling tones is usually carried out by dedicated circuitry in the transceiver, based on purpose-designed integrated circuits.

There are three basic systems of signalling involving transmission of either:

- continuous tones
- sequential or burst tones
- combinations of tones

These are some examples of tone signalling in PMR: ·

- Continuous tone controlled signalling system (CTCSS)

CTCSS (also known as 'private line', 'tone lock' etc.) is the most common method of signalling where several user groups wish to use a single channel without each group hearing the others. A tone encoder/decoder fitted to the transceiver modulates all transmissions at a low level (300Hz deviation in FM [17]) with a specific low frequency tone in addition to the normal audio, each user group working with their own tone. Receiver output is monitored by the tone decoder which opens the receiver squelch only when incoming transmissions are modulated with the required tone. 32 standard tones are available (Annex B) ranging from 32Hz to 250.3Hz, which are readily filtered out from the received audio.

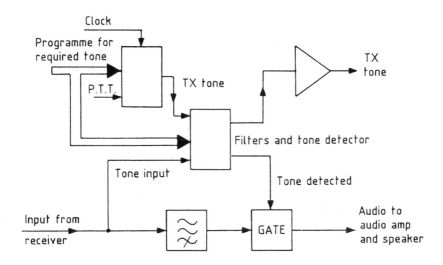

Fig. 5.13 CTCSS dencoder arrangement

Some form of monitoring of the channel is necessary to prevent users accidentally keying over a transmission in progress, alternatively a 'lockout' circuit in the transmitter press-to-talk circuitry could be included which prevents a transmission being made if the channel is already busy with a transmission carrying another user's tone. RA will also licence a dispatcher-controlled talk-through facility if the system operates with CTCSS since it can be arranged via a decoder at the base station that only mobiles in the user group transmitting the required CTCSS tone will operate the talkthrough facility [18].

- Digitally controlled squelch (DCS)

In DCS systems (also known as 'digital private line' etc), transmissions carry a continuous low level low bit rate digital signal consisting essentially of a repetitive codeword representing a specific 3-digit number. At the receiver a decoder responds only to the incoming codeword representing the number programmed into the decoder. The advantage over CTCSS is that there are far more individual codes possible (104 recommended) and that the system response to an incoming transmission is very fast. The DCS encoder may also send an end-of-transmission code to close receiver squelch, eliminating any squelch tail.

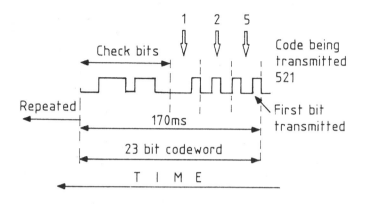

Fig. 5.14   DCS encoding

- Sequential tone signalling

There are many PMR systems in which the dispatcher needs to call specific mobiles, for example in a fleet of delivery vehicles, as opposed to mainly 'broadcasting' calls as in the case of a taxi operation where many dispatcher messages need to be heard by all mobiles. Selective calling [19] or selcall could be achieved by sending a tone or sequence of tones within the audio passband of the transceiver, the number of tones available for each tone 'slot', and the number of slots determining the capacity of the system. Several standard formats have been published which involve the transmission of a sequence of tones of standard duration, each tone representing either a digit or an instruction. Two-tone signalling has been in use for some time. However, the higher capacity of five-tone systems has generated greater interest and application in recent years (Annex C).

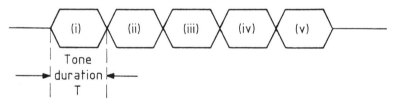

(Each tone 'slot'may be occupied by any one of the tones from the tone set (Annex A)

Fig. 5.15   Sequential tone signalling

In a basic selcall system, each mobile in the fleet is fitted with a decoder preprogrammed with the code specific to that mobile: for example a 20 vehicle fleet could have the numbers 12301 to 12320. The dispatcher uses an encoder at the base station to select and transmit the five-tone code of the mobile being called. The mobile is usually muted until the correct incoming selcall code is received so that drivers/operators hear only calls from the dispatcher directed to them. The mobile may also indicate the 'called' condition by an LED or audio tones to alert the driver and additionally make an auto-acknowledge or revertive transmission of either a single tone or the mobile's own five tone code to confirm a successful contact to the dispatcher. The standard formats include group calling - in this example 1230G would call all mobiles 12301 to 12309 - and repeat tones, so that mobile 11 is called as 1231R which avoids two successive tone slots carrying the same tone.

The mobile may also use its selective calling facility to indicate a request for attention to the dispatcher, by making a short transmission of its own 5-tone code. In such a system, known as vehicle identification (VI) or automatic number indication (ANI), a decoder at the base station displays incoming selcall numbers, which could be stacked in a busy system or displayed to the dispatcher via a VDU, offering scope for sophisticated methods of control.

Mobile status may be established by providing a facility on the transceiver for adding a further digit or digits to the tone sequence reverted which may be set by the user. The dispatcher then polls the mobile to check the status which would be displayed with the mobile identity so that for example the progress of a service technician round the day's list of calls could be established even if the vehicle were unattended.

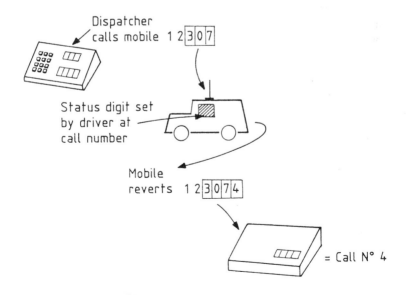

Dispatcher
calls mobile 1 2 3 0 7

Status digit set
by driver at
call number

Mobile
reverts 1 2 3 0 7 4

= Call N° 4

Fig. 5.16 Outgoing selcall, with status revert and display

Where there may be urgent need for attention, in a security vehicle or one-man bus situation for example a 'panic' button on the mobile could initiate a specific tone sequence which could be arranged to override the stack displayed to the dispatcher.

Sequential tone signalling has also been used extensively for triggering pagers, either directly (overlay paging) or from the revertive transmission of a mobile (secondary paging) which alerts the driver when away from the vehicle (Fig. 5.17).

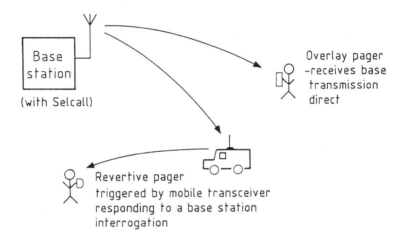

Fig. 5.17 Overlay and revertive paging in PMR

Another application of sequential tone signalling from the mobile is to obtain access to a talkthrough base station or to switch such a base into talkthrough. Equipment is also available which enables a mobile to access a company's private automatic exchange (PABX) and make calls to office extensions via an interface to the base equipment, the last digit or two digits indicating the extension required to a directory dialling system in the interface. Time on-air with interconnects of this type is however limited under licence conditions [20] to prevent excessive use of the radio channel (telephone conversations being notoriously longer than an average two-way radio contact).

- Dual tone multi-frequency (DTMF) signalling

DTMF (or 'touch tone') encoding uses a matrix of row and column frequencies to represent a standard 12 key telephone keypad (Annex D). Each digit is sent by simultaneous transmission of the two appropriate frequencies, one low and one high in range, which are between 697Hz and 1477Hz and well within the audio range of a PMR unit. Again, one application of DTMF signalling is for a mobile to access a PABX with a combined DTMF keypad and fist microphone on the transceiver providing a convenient method of dialling in.

5.7 Remote control

Transmissions made from elevated hilltop sites or tall tower blocks provide wide areas of coverage and such locations are popular with many of the larger users of PMR for siting of base equipment. Remote control of transmitter - receiver functions and the audio path to and from the user's premises is commonly achieved via a BT leased line ('private wire') dedicated to the user. Methods of control depend on the type of line made available (either a two- or four-wire system) and over ranges of a few km a dc path has been possible. In this latter case, the landline control unit at the user's premises supplies a dc voltage, typically at a nominal 50 volts, for keying and other switching functions [21].

Current BT practice is to provide an ac path, which requires some form of tone signalling. There is no standard method and techniques include sending one or two continuous audio tones generated at the control point and detected at the remote control panel for transmitter keying, other switching functions such as talk-through on/off and to indicate the status of the line (e.g. loss of normally continuous tone equals line failure). Another possibility is to send continuous FSK tones, for example 2898Hz and 3011Hz, with specific mark-space ratios to initiate the required function at the remote end. A more sophisticated method would be to send instructions via FFSK data bursts.

Where it can be shown that no BT line is available [22] RA will allow the base station to operate in talkthrough, with a static mobile working from a mains/dc power unit at the dispatcher position, an arrangement termed 'reverse frequency working'.

Remote operation of base equipment is also permitted using microwave link equipment at 1450MHz to 1690MHz (Annex A), the links carrying transmitted and received information and keying signals within the audio range in similar form to those described above for a landline.

5.8 Common base station or repeater systems

Repeaters [23] are a logical extension of the principle of a talkthrough base with CTCSS, and are very efficient in the use of the spectrum. A repeater is usually licensed by a company in the PMR industry who in turn authorise the use of the repeater (for an annual rental) by several different groups of users, each group having its own CTCSS tone. The CBS system through a bank of CTCSS decoders detects the tone on an incoming transmission and re-transmits both incoming audio and, usually, the same CTCSS tone, i.e. Fig. 5.18.

As before decoders in the mobiles ensure that transmissions through the repeater are only heard by the relevant user group. Monitoring facilities are also required on the mobiles or PTT lockout as before. A transmit timer circuit is also desirable as would be timers on the base station to ensure that users cannot 'hog' the system. Dispatcher units or 'trigger bases' at the various users' premises are again typically a static mobile with a mains/dc power supply. RA require trigger bases to operate with a directional antenna (usually a Yagi) aimed at the repeater site and with transmitter power level determined by the signal received from the trigger base at the repeater.

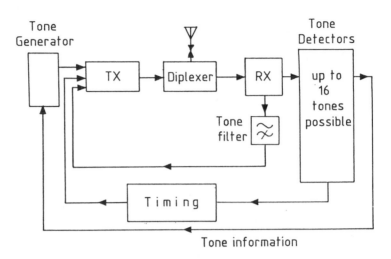

Fig. 5.18 The common base station arrangement

5.9   Data transmission in PMR

The rate of transmission of information by digital methods is potentially far greater than using speech, particularly in situations where hard copy (printer, fax) or graphics (with a CRT or LCD panel) are used. Data transmission also gives direct access to computer services such as data bases, reduces message handling time, and increases accuracy (in conjunction with error correcting protocols). Shorter transmission times can significantly increase channel efficiency, providing the opportunity for increasing the number of mobiles operating through a single channel or reducing channel loading. For example, details of a customer name and address and other particulars relevant to a call may take some minutes to pass using voice: the same information in a data burst could occupy the system for only a few seconds. In general data systems are 5 - 20 times faster than voice [24,31]. There is also the spin-off that digital transmissions offer privacy, particularly against the casual eavesdropper. Since all transactions pass through a computer, suitable software would offer considerable scope for reducing paperwork particularly in conjunction with direct input of data, for example from bar code readers or portable data gathering terminals.

Selective calling and DTMF are possible methods of sending data and have been used as such, however data rates are relatively slow although the long duration of tones makes it possible to decode them reliably even in very poor signal conditions.

Digital transmissions over mobile radio channels have in the past tended to use specialised modems but methods developed in recent years for data transmission over the PSTN (CCITT V series) are now more common. 1200 bits/s FFSK using $0 = 1800$Hz and $1 = 1200$Hz has proved to be consistently reliable in mobile radio [25], however designers are still basically free to use any signalling method provided that the transmission is contained within the allocated part of the channel without spurious sidebands. In practice data transmission rates over a 12.5kHz channel are therefore limited to 2400 or 4800 bits/s.

Again, protocols are at the designer's choice, taking into account the unique characteristics of the mobile radio channel compared with the PSTN. The integrity of data presented to the user is influenced by the presence of multipath (Rayleigh) fading and dropouts caused by bridges, high buildings and other ground features [26]. The duration of fades may be from fractions of a second to several seconds with rates up to several fades/second at high speeds, which require techniques of error detection and correction. Codes which require few redundant error-correction bits offer higher throughput but may be too weak to correct errors, with the resulting requests for repeated transmissions (ARQ). Very powerful forward error correction (FEC) codes have considerable redundancy and hence reduce throughput so that a balance is required taking into account the nature of the data and the channel characteristics.

In the 1200 bits/s FFSK EEA/DTI data format [27], code words are 64 bits in length with 48 information bits and 16 check bits used for error detection.

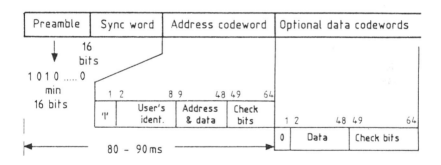

Fig. 5.19   The UK standard FFSK data format

The address code word contains user's identity, address and check bits; there may also be data (such as status reporting) in the address code word in short messages. In longer messages the data code word which follows an address code word contains data and check bits and is basically as long as is required.

Since FFSK transmissions are originated by keying two audio frequencies, these may be input to the transmitter via the normal audio circuitry provided that the rate of signalling is not higher than 2400 bits/s. At higher data rates, e.g. 4800 bits/s, direct frequency modulation by data at baseband to produce FSK, rather than via FSK tones, is essential (phase modulation is not capable of holding the carrier at a fixed deviation). Spreading of the transmission at these speeds is avoided by using reduced deviation of around 1.2kHz. Equipment modified for data transmission by direct modulation has to be resubmitted for testing by RA in the new configuration.

Fig. 5.20 shows a data system typical of that required for taxi dispatching or a TV repair company where requests for calls are made on an ongoing basis during the working day.

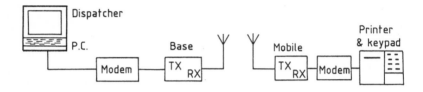

Fig. 5.20 The basic data transmission arrangement

Customer details are logged into the system which allocates a vehicle for the call. The computer interrogates the mobile to establish contact and on receipt of an acknowledgement sends the data as a burst transmission. If an ARQ protocol is in use, the mobile checks received data for errors, requesting a whole or partial repeat if required before producing a printout or displayed text. This could include customer name and address and any further relevant details. Acceptance of the work, arrival at customers' premises, completion, request for the next assignment and so on may be signalled from the mobile terminal using dedicated keys.

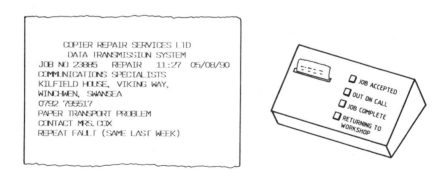

Fig. 5.21 A basic mobile data terminal and printout

Smaller systems of this type with less than 100 mobiles could carry both voice and data on a single channel. In larger or busier systems a separate voice channel would be essential to take full advantage of the increased capacity of a data transmission system.

An important consideration in a mobile data system is that the requirement for mobiles to make transmissions occurs at essentially random intervals giving rise to the possibility of collisions and loss of data [28]. Some possible solutions include:

- allowing the mobiles to operate a system of uncontrolled contention (pure ALOHA) with perhaps error correcting codes to improve the quality of corrupted data.

- polling each mobile data terminal (MDT) in turn (which is wasteful of time and would severely reduce throughput of data).

- carrier sensing, where the base transmits a busy signal during periods of incoming data from a mobile, signalling other MDT's to wait [29]. Once the channel is seen to clear mobiles contend for the channel: if there is any collision, mobiles wait for a random interval before retrying.

- use signalling formats by which the base can indicate time slots in which mobiles are invited to transmit (slotted ALOHA): again if there is a collision in a particular time slot, mobiles wait a random number of time slots before retrying.

The potential for improving throughput of traffic in a dispatcher system is considerable. For example the number of mobiles controlled by one dispatcher may increase by a factor of two or three [30]. Larger taxi operators are able to pass information to the cabs in 30 - 40 secs from receiving the customer's call instead of 10 - 15 mins [31,32].

Mobile data is also finding applications in goods handling, particularly in for example warehouse and dockside situations where loads are assembled in conjunction with manifests held on computer.

Fig. 5.22 illustrates a larger user's data transmission system which could be typical of one of the emergency services.

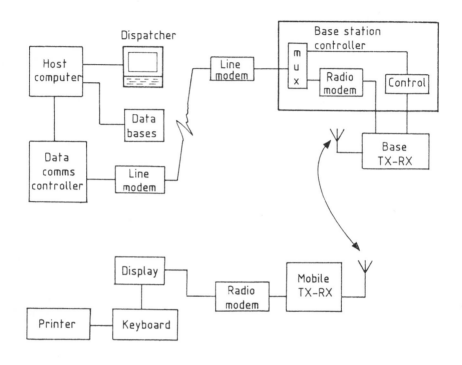

Fig. 5.22 A full data communications facility

There may be one or more data channels and a separate voice channel shared perhaps with data traffic. Access to data bases will form an important part of such a system and require an MDT which includes QWERTY keyboard, display using either CRT or LCD and probably a printer in the mobile. The quality and detail of information displayed in the mobile could be enhanced by onboard intelligence and mass storage devices called up by the host computer [33].

5.10 Area coverage techniques

Coverage from a single site is limited by factors such as local topography, frequency band in use, base station aerial elevation and gain, and base and mobile transceiver parameters. Many users with wide operational areas require coverage which can only be achieved by base stations on two or more sites; examples are railways, county authorities, emergency services, and the Fuel and Power industries. Another requirement for multiple site working is where gaps in main site coverage caused by ground features can be filled by additional base stations.

Working through more than one base station can be achieved by simply operating each base station on a different channel. Mobile operators then select the required channel according to their location and with knowledge of the coverage of each base station.

- Synchronous and quasi-synchronous operation ('Simulcast')

In systems having wide areas of coverage, and especially in those with gap-filling requirements, it would seem logical to operate all sites on the same channel and, to ensure that all mobiles hear all transmissions, to transmit simultaneously from all sites. In principle, transmitters should be locked in frequency - i.e. synchronous - to avoid a heterodyne effect in areas of overlapping coverage. This will however produce a stationary pattern of constructive and destructive interference in such areas, much the same as multipath interference from a single site, where deep nulls will give rise to many points where no reception is possible when the mobile is not moving. The further apart the sites are, the greater the area affected. Doppler shifts on the transmissions from each site received by a mobile whilst in motion will differ and may give beats up to 80 - 100Hz [34]. Synchronous operation can be achieved [35] but the requirement for identical transmissions from all sites can only be met by specialised and expensive systems.

Quasi-synchronous operation is a more practical solution. Transmitters at each site are offset in frequency by up to 30 - 40Hz (0.5 - 4Hz for AM systems, 5 - 40Hz for FM [36]) and although this produces beat notes when two transmissions are received, the note is well below the mobile audio response: constructive and destructive points of interference will also move around the area of mutual coverage and the effect on mobile communications will be to give a degree of flutter on the transmission which is not judged to be too disruptive to good communications. Quasi-synchronous operation does not require that sites are linked at RF to provide a reference for transmitter synchronism, and carefully matched but independent frequency sources at each site are sufficient.

There is a stringent requirement in synchronous and quasi-synchronous systems for the paths from the control point to each transmitter site to be equalised to prevent distortion on received signals. Modulation must also be identical in depth/deviation so that landlines must be matched in gain/loss, phase and frequency response. Such stringent requirements in a part of the system outside the control of the system operator may preclude the use of landlines in favour of microwave links which form part of the user's own infrastructure. Provided that sites are linked via the control point, talkthrough is possible from one site to another and all mobiles hear all others. Data transmission however is a particular problem due to distortion and flutter in the overlapping coverage areas [37].

- Multi-channel operation

Many users have avoided the complexity and reception problems of quasi-synchronous systems by licensing several channels, each covering an area of interest. Sites are linked to one or more control points and may be interconnected, whilst mobiles fitted with multi-channel transceivers simply select the appropriate channel for their area of operations. This is however wasteful of spectrum and requires some skill on the part of the mobile operator in matching the necessary channel with his present location, although systems of this type are in general use largely because of their relative simplicity. They were also, until recently, typical of the larger 'message handling' operations (for example Aircall) which provided independent mobile users with a dispatcher having access to the PSTN who would relay messages to and from the mobile. Message handling has however largely been displaced by cellular telephones.

- Voting in wide area coverage

Voting is an efficient way of achieving extended coverage since only a single channel is required. The required area of operation is covered by as many base stations as required, however only one base station transmits at any time so that mobiles in areas of overlapping coverage do not experience the interference problems encountered in synchronous systems. Sites are linked back to a control point via a voting system whose role is to evaluate incoming transmissions from all sites on for example a signal/noise basis either at the output of each base station or at the control point to which the base stations are linked. A decision is then made on each transmission received as to which of the available outputs offers the best service and this base station is connected to the dispatcher. On completin of the transaction, the system resets awaiting the next call.

If the dispatcher is calling a particular mobile, selective calling for that mobile is transmitted from each site and voting takes place on the revertive signal from the mobile. Talk-through could be a problem however if the mobiles concerned are not in the coverage area of the same transmitter.

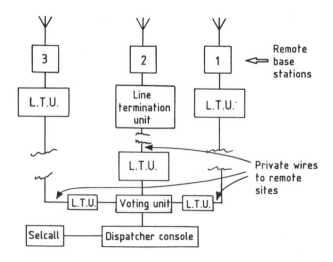

Fig. 5.23 A voting arrangement

5.11 Conclusion

Private mobile radio is a thriving if congested service, with users in all parts of industry and systems ranging from the very small to some operating thousands of mobiles. Areas of operation may be a few hundred square yards to two or three counties.

Well proven techniques are in place enabling users of PMR and those who service the industry to improve the performance of their systems without placing extra demands on the spectrum, with improved forms of signalling and, particularly, data transmission having a significant role [38].

5.12 References

1. DTI: 'UK Table of radio frequency allocations' (HMSO, 1987)

2. MACARIO, R.C.V.: 'Mobile radio telephones in the UK' (Glentop Press, London, 1988), p. 129

3. Radiocommunications Agency newsletter, No. 2, Aug 1990

4. DTI: 'Annual report 1988/89', p. 43 (Central Office of Information)

5. DTI: 'Sharing of private mobile radio channels', Fact Sheet BR53

6. DTI: 'Angle modulated VHF and UHF radio equipment for use at base and mobile stations in the PMR service', MPT 1326, Nov 1985, p.13

7. DTI: 'Angle modulated VHF and UHF radio equipment for use at base and mobile stations in the PMR service', MPT 1301, Jan 1983

8. DTI: 'Amplitude modulated VHF radio equipment for use at base and mobile stations in the PMR service', MPT 1302, Mar 1977

9. DTI: 'Code of practice for radio site engineering', MPT 1331, April 1987

10. DTI: 'Legal requirements covering the installation of radio apparatus and other apparatus generating radio frequency emissions', MPT 1367, July 1988

11. DTI Leaflet: Kenley radio technology laboratory, 'Radio equipment type - approval services'

12. DTI: 'Angle modulated VHF and UHF radio equipment for use at base and mobile stations in the PMR service', MPT 1326, Nov 1985, pp. 10 - 18

13. DTI: 'Angle modulated VHF and UHF radio equipment for use at base and mobile stations in the PMR service', MPT 1326, Nov 1985, pp.19 - 25

14. BELCHER, R., FITCH, M., OGLEY, D., and VARRAL, G.: 'Mobile radio servicing handbook' (Heinemann, London, 1989) pp. 78 - 79

15. GRIFFITHS, J.: 'Radio wave propagation and antennas' (Prentice-Hall International, 1987)

16. Electronic Engineering Association: 'Handbook for Private Mobile Radio Users', Feb 1986

17. DTI: 'Continuous tone controlled signalling system (CTCSS) for use in the PMR service', MPT 1306, Jan 1978

18. DTI: 'Private mobile radio licence application guidance notes', p. 6

19. DTI: 'Code of practice for selective signalling for use in the PMR service', MPT 1315, Jan 1981

20. DTI: 'Private mobile radio licence application guidance notes', p. 6

21. DAVIES, J.: 'Private mobile radio, a practical guide' (Heinemann, London, 1987), pp. 70 - 73

22. DTI: 'Private mobile radio licence application guidance notes', p. 7

23. DTI: 'Code of practice for repeater operation at communal sites', MPT 1351, June 1988

24. STANNARD, C.: 'Mobile data transmission - the missing link', Mobile Telecommunications News, July/Aug 1988

25. MACARIO, R.C.V.: 'Mobile radio telephones in the UK' (Glentop Press, London, 1988), p. 114

26. PARSONS, J.D., and GARDINER, J.G.: 'Mobile communication systems' (Blackie, London, 1989), Chaps 2 and 3

27. DTI: 'Code of practice for transmission of digital information over PMR systems', MPT 1317, Apr 1981

28. KLEINROCK, L.: 'Queuing systems', Vol. 2. (John Wiley, New York, 1976)

29. MORRIS, J.: 'MDT's give mobiles direct access to digital computer data bases', Mobile Radio Technology, Dec 1983

30. JOHNSON, P.D.: 'A major user's view of mobile radio today', MRUA Conference, Oxford, 1989

31. MORANT, A.: 'Terminal trafficking', Mobile Business News, May, 1990

32. MORANT, A.: 'Profiting from a better service', Quarterly Journal of the National Federation of Taxicab Associations, Summer 1989

33. WORSELL, M., and WHITEHEAD, P.: 'An integrated command, control and data network for West Midlands Fire Service', MRUA Conference, Oxford, 1989

34. ATACK, R.: 'Simulcasting without tears', Mobile Telecommunications News, Apr - July 1990

35. HOLBECHE, R.J.: 'Land mobile radio systems' (Peter Peregrinus, London, 1985), Chapter 7

36. DAVIES, J.: 'Private mobile radio, a practical guide' (Heinemann, London, 1987), p. 66

37. MABEY, P.J.: 'Data systems', in HOLBECHE, R.J.(Ed.): 'Land mobile radio systems' (Peter Peregrinus, London, 1985)

38. Radiocommunications Agency: 'Annual Report 1989/90', p. 8, (HMSO)

Annex A

PRIVATE MOBILE RADIO BANDS (1990)

Band	Type of Service	Frequency Band (MHz)		ERP (watts)
VHF				
Low	Two frequency simplex			
	Mobile transmit	71.5125	- 72.7875	25
		76.9625	- 77.50	
	Base transmit	85.0125	- 86.2875	
		86.9625	- 87.50	
	Single frequency simplex	86.30	- 86.70	5
Mid	Two frequency simplex			
	Mobile transmit	105.00625	- 107.89375	25
	Base transmit	138.00625	- 140.99375	
	JRCBase	139.51875	- 140.48125	
	Mobile	148.01875	- 148.98125	25
High	Two frequency simplex			
	Base transmit	165.05	- 168.25	25
	Mobile transmit	169.85	- 173.05	
	Single frequency simplex	168.95	- 169.85	5

(169.4 to 169.8 to be vacated by 31.12.91)

Additional channels, formerly used by Radiophone System 4, to be allocated to various PMR services:

Mobile transmit	158.5375	-	159.8120
Base transmit	163.0375	-	164.3125

SERVICE PROVIDER CHANNELS (formerly Message Handling)

Two frequency simplex				
Mobile transmit	159.9375	-	160.5375	25
Base transmit	164.4375	-	165.0375	

UHF

'down band'   (available for use in London, Birmingham, Manchester, Liverpool, Aberdeen, Newcastle, Middlesborough, Leeds, Bradford, Halifax, Sheffield, Nottingham, Derby, Leicester, Preston, Edinburgh and Glasgow areas)

Two frequency simplex					
Mobile transmit	425.025	-	425.475		25
	425.525	-	428.975		
Base transmit	445.525	-	445.975		
	440.025	-	443.475		
Single frequency simplex	446.025	-	446.475		5

(available for use in London)

Two frequency simplex					
Mobile transmit	431.00625	-	431.99375		25
Base transmit	448.00625	-	448.99375		

'up band'

Two frequency simplex					
Base transmit	435.025	-	453.975		25
	456.00	-	456.975		
Mobile transmit	459.525	-	460.475		
	461.50	-	462.475		
Single frequency simplex within the above ranges					5

FIXED POINT-TO-POINT LINKS

Up to 36 telephone channels				
or equivalent in FDM/FM	1450	-	1462.5	600
3 to 8 channels	1467.5	-	1472.5	200
1 + 1 channels	1472.5	-	1479	100
	1503.5	-	1506.5	

ANNEX B

CTCSS EIA STANDARD TONE FREQUENCIES

EIA Group A	
Number	Frequency (Hz)
1	67.0
2	77.0
3	88.5
4	107.2
5	114.8
6	123.0
7	131.8
8	141.3
9	151.4
10	162.2
11	173.8
12	186.2
13	203.5
14	218.1
15	233.6
16	250.3

EIA Group B	
Number	Frequency (Hz)
1	71.9
2	82.5
3	94.8
4	103.5
5	110.9
6	118.8
7	127.3
8	135.5
9	146.2
10	156.7
11	167.9
12	179.9
13	192.8
14	210.7
15	225.7
16	241.8

ANNEX C

SELECTIVE TONE FREQUENCIES

Digit	Mod. ZVEI	tone frequencies (Hz) EEA/CCIR	EIA
1	970	1124	741
2	1060	1197	882
3	1160	1275	1023
4	1270	1358	1164
5	1400	1446	1305
6	1530	1540	1446
7	1670	1640	1587
8	1830	1747	1728
9	2000	1860	1869
0	2200	1981	600
R	2400	2110	459
G	825	1055	2151

Tone Duration

ZVEI   70 mS
CCIR   100 mS
EIA   33 mS

ANNEX D

DTMF TONES (Hz)

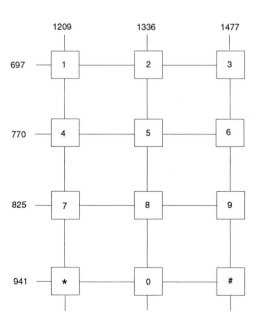

Chapter 6

Trunked mobile radio systems

David E A Britland

6.1 Introduction

The growth in demand for frequencies for land mobile radio services has been steadily increasing over the past few years. The advent of mass cellular radio usage has done little to diminish the requirement for spectrum mainly because Private Mobile Radio (PMR) in general addresses a different group of users. The ever increasing demand has been difficult to satisfy with a finite resource. The UK DTI have sought to alleviate this problem by allocating Band III sub-band 2 frequencies to public trunked radio systems. Trunking makes greater use of a channel, but leaves the user less aware of the congestion on that channel. Users share a pool of channels and are only allocated a channel when they need to make a call. In practice not all users wish to make a call at the same time and trunking theory is based on the probability that there will be a free channel when required. As the number of channels in a group exceeds three, the gain from trunking starts to become significant. The telephone networks have been using trunking theory and practice for a great many years, but it has only recently become economically possible in radio systems with the advent of microprocessors and low noise synthesisers. Twenty years ago trunked radio systems were in their infancy and relied on banks of switched crystals to change channel. Ten years ago, in order to support 5 tone and DTMF signalling together with channel frequency crystals, portable radios resembled walking sticks with a battery on the bottom. Now it is possible to shrink all the intelligent control into a personal radio by use of synthesisers, microprocessors and digital signal processors (DSP's).

6.2 Benefits of a trunked system

In a trunked scheme a larger number of mobiles can be accommodated for a given grade of service than could be supported on an equivalent number of single channel systems. Conversely the same number of mobiles on single channel systems could be given a better grade of service on a trunked system. Or, the same number of mobiles could be given the same grade of service on fewer channels, thus saving spectrum.

The reliability of a trunked system is very much greater. The loss of a single base station merely degrades the service. To a single channel user this could mean total loss of communication.

Privacy is inherent in a trunked system because of the use of selective signalling. The degree of privacy is controllable from individual calls through to group and fleet calls.

Usually it is relatively simple to expand trunked schemes by the addition of extra channels. Because the mobiles are frequency agile no alteration is normally required.

As far as a mobile operator is concerned, a trunked system need be no more complicated than using a single channel radio. The radio and the system can set up calls automatically. More complex keypad mobiles can be added for those who require multiple address calls. Wide area coverage is possible by suitable siting of base stations. The system controller will then select the optimum site for communication. The cost of the infrastructure can be shared between a large number of users. This may be offered through a service provider who would charge a monthly fee. Capital cost and the need for individual licences are also removed from the small user.

6.3 System considerations

Channels must be allocated with a minimum of delay. In the Band III specifications 1200 baud fast frequency shift keying (FFSK) has been chosen as a compromise between speed and reliability within the 12.5 kHz channel separation. The TACS cellular system employs 8 kbit/s direct modulation. (The effective channel separation of 50 kHz can accommodate this). Systems using 3600 and 4800 bit/s direct modulation are in use, mainly in the USA.

The protocol must ensure utmost reliability of correct connection within a harsh environment. The signalling telegrams themselves have powerful error detection capability so it is unlikely that significant falsings will occur. However, many telegrams are likely to be rejected in bad signal conditions and therefore the protocol must be predicated on acknowledgements and retries. A balance between no retries and an excessive number must be struck which does not overload the control channel.

To some extent, the above problem can be alleviated by measuring the ongoing error rate of the control channel and rejecting it and hunting for a new one should the errors become too bad. This is a central part of the MPT 1343 strategy (section 9). This assumes reciprocal operation which is not always the case, particularly with interference present. Nor does it allow for the case of corrupted codewords due to simultaneous access attempts by mobiles.

A significant problem which can occur in trunking systems which operate in a two frequency simplex mode is that sometimes signalling may be missed due to the intended recipient being in the transmit state. For instance, should the system require all mobiles to clear down due to an emergency or even a simple time out, those mobiles already transmitting will continue to do so. Various strategies have been employed to minimise this problem which if ignored can lead to quite serious degradation in the grade of service.

The inclusion of queuing in a system design has two main advantages. If a system has become very busy the only means of access to a traffic channel, if there is no queuing, is by continuous retries. This may be accomplished manually or automatically or a combination of both. These retries will tend to block the control channel and at the moment a traffic channel becomes available contention could be so great as to require time to re-allocate it. This reduces the efficiency of the system and generates a lot of frustration to the user. Secondly, the user will have a more favourable perception of the system when he obtains a queued message. Also emergency and priority calls can be more readily processed.

Operation of a trunked radio system need not be limited to any specific frequency band or channel separation. However, different signalling methods are better suited to the frequency at which it operates. For example, in the low VHF band ignition interference is prevalent so a signalling method which incorporates single bit error correction is likely to be more efficient. At 900MHz the main cause of problem is likely to be rapid Rayleigh fading so short telegrams (in time, not necessarily content) with burst error correction capability would be preferable.

6.4 Technical considerations

As previously stated, various forms of signalling have and are being used in trunking systems. The method currently being adopted in Europe is the 1200 baud FFSK indirect modulation scheme. However, there are several different approaches to the format of the telegrams. In the UK the MPT 1317 format has been chosen and is being widely adopted in Europe. In Scandinavia and France the Nordic Mobile Telephone (NMT) format using forward error correction (Hagelbarger code) has been used. However, in France an alternative code PAA 1382 is now available which is very similar to MPT 1317.

In brief, the MPT 1317 telegram is made up from a bit synchronisation sequence, a word synchronisation sequence, a 48 bit data field and a 16 bit cyclic redundancy check (CRC).

BIT SYNC	WORD SYNC	DATA	CRC
16 bits	16 bits	48 bits	16 bits

├——— codeword ———┤

The 64 bits comprising the data and CRC are known as a codeword. The MPT 1317 and PAA 1382 codewords are identical. The difference between the standards lies in the bit sync. and the word sync. The PAA 1382 being a slightly newer standard has optimised the auto correlation function of the synchronisation patterns. This is of use if error correction action is to be undertaken. However, for signalling rather than data transmission, error correction is of less importance because the protocol allows for retries and acknowledgement. Again a balance has to be struck as to whether the overhead of redundancy of error correction in all signalling is greater or less than the retries of error detected packets. An additional argument can be made that if the channel is unable to support simple short telegrams then it will also be unable to support speech. This being said the MPT 1317 format does allow for minimal error correction [1], Appendix 2, within the codeword of:

(i) Detect all odd numbers of errors, any 5 random errors and any error burst up to length 16, or

(ii) Correct any one error and detect any 4 errors and any error burst up to length 11, or

(iii) Correct up to any 2 errors, and detect any 3 errors and any error burst up to length 4, or

(iv) Correct any single error burst up to length 5.

The more error correction which is performed the greater the chance of falsing. In the case of NMT and Radiocom 2000, which use error correction, the falsings are controlled by careful use of a handshaking protocol.

In Band III systems a modified form of the above telegram is used on the continuous forward control channel. This is an amalgamation of two telegrams to generate a time slot of 106.7 mS. After the initial start up, the bit and frame synchronisation for address code words becomes part of the previous codeword. That is, the last 32 bits of the Control Channel System Codeword (CCSC) must always compute to be the same. This is achieved by freely allocating bits 2 to 16 for a system identity (bit 1 is always zero) and then computing the next 16 bits together with bits 33 to 48 which are reversals to generate a CRC which is the same as the frame synchronisation pattern. In this way it is possible to do 2 CRC checks on each time slot, i.e. the CCSC and the address codeword. The detailed codewords, channel discipline, random access procedures, registration procedures, etc., are to be found in MPT 1327.

The system interface requirements for use in commercial trunked networks in Band III sub-band 2 can be found in MPT 1343.

A trunking system may be either traffic channel or control channel limited. If conversations are excessively long then the limitation is the availability of traffic channels. Unlike cellular telephones most trunking systems do not charge for airtime and therefore usually fit call timeouts. A one minute timeout is quite normal. Conversely if calls are very short then the limitation becomes the control channel. The assumptions used in MPT 1318 for the grade of service of 20 second wait and call times have been substantially met. However, the average call length is now accepted to be approaching 35 seconds.

6.5 Control channel

It has been shown [2] that a framed slotted Aloha protocol can allow 1/e random accesses on the control channel in the mobile to base direction (return control channel). Because the base to mobile (forward) direction is centrally controlled, 100% may be used but some capacity is lost due to system overhead. At inputs greater than 37% (1/e) the system would become unstable due to contention unless further precautions are taken. By use of dynamically varying frame length and sub-division of the population it is possible to maintain stability and also increase the throughput to 43%. These figures are theoretical and have been computer simulated. In the field they are usually better due to the use of multi-receiver sites and the effect of FM capture. In some system they can be worse due to local interference. That being so, they are a useful base on which to calculate control channel capacity. Using the above information it is possible to estimate the number of calls per hour that a single control channel can support. In the case of MPT 1327 the slot length is 106.7 ms.

Therefore there are $\dfrac{1000}{106.7}$ x 60 x 60 slots/hr

$\qquad$ = $\quad$ 33750 $\quad$ slots/hr

of which 37% are useful $\quad$ = 0.37 x 33750

$\qquad\qquad\qquad$ = 12415 slots/hr

A mobile-to-mobile call on average will require two reverse channel slots (A landline dispatcher will require 1, an access to PABX will require 2, a queued system at least 3).

However, it is usual because of the hostile mobile radio environment, that only 85% of the calls will access with no retries caused by fading, interference, etc.

Therefore the control channel on average can set up:-

$\dfrac{12415}{2}$ x 0.85 calls/hr

$\qquad$ = 5276 calls/hr

6.6 Traffic channels

Trunking theory is based on statistical probabilities. It is assumed that the requests for service follow a Poisson or exponential distribution. The theorem adopted for mobile radio is the Erlang C formula [3] and is based on the following assumptions:

-   an infinite number of users
-   random interval between calls
-   random call duration times
-   negligible call set up times
-   first in, first out (FiFo) queuing

A practical radio system does not meet all these criteria but the Erlang C can provide approximate performance figures.

The measure of congestion (grade of service) is the means by which a system may be quantified. This is the probability that a user will experience a specified delay in the busy period (busy hour). The delay gets worse as channel loading increases and saturation occurs when the specified delay is consistently exceeded. Being a probability of delay the lower the value the more acceptable is the system to a user.

In order to calculate the grade of service it is necessary to know or estimate the expected acceptable delay and also the channel holding time. This is the total period for which a channel is engaged by the user for the purpose of sending and receiving speech or data. This does not include the call set up/clearing down times. When planning a system the holding time is the mean value per mobile in the busy hour of an ordinary day. In addition it is necessary to estimate the number of calls the average mobile will make in the busy hour. By these means it is then possible to estimate the traffic capacity and the number of channels required to support a given number of mobiles with the required grade of service.

As an example using Table 1 in MPT 1318 [4] for a 20 channel system where:-

The mean channel holding time is 20 s.
The waiting time in the busy hour is 20 s.
Then for a 5% grade of service each traffic channel exhibits a capacity of 0.885 Erlang and mean waiting time of 4.3 s per call. This excludes signalling time, the addition of which will make the mean waiting time approximately = 5 s.

The 4.3s delay is shown in Fig. 6.1, reading off the 20 channel system graph at the 0.885 Erlang channel loading point.

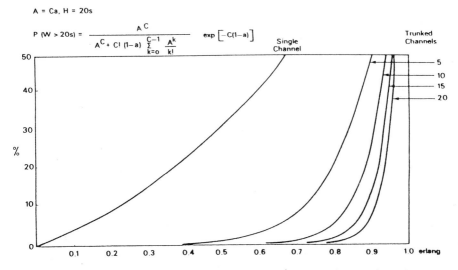

Fig. 6.1   Probability of delay greater than 20 seconds as a function of channel load (a) for given number of channels (C) and a holding time (H) of 20 seconds

Assuming each mobile makes one call/hr., then each channel will support 159 mobiles. The 20 channels will therefore support 3185 mobiles or calls per hour. (One call per hour per mobile).

It can therefore be seen that a single control channel can easily support a 20 traffic channel system as only 60% of the available capacity is utilised. It can be shown that a single control channel may support up to 32 traffic channels before it becomes the limiting factor in the system at this grade of service. This applies to single trunked system controllers (TSCs) which may serve relatively small areas. If the system is to be expanded for wide area coverage with TSC interconnection further demands may be made on the control channel for registration, forced to move messages, etc., then it is advisable to limit the number of traffic channels per control channel to allow for the extra signalling which will be required.

In wide area systems where one base station site is unable to give the required coverage, control channels will have to be associated with each site. In order to save on spectrum it is possible to time-division-multiplex the same control channel frequency to broadcast sequentially at each site. This decreases the capacity of the control channel by approximately 15% due to start up and identification sequences required to synchronise mobiles. The grade of service as perceived by the user is also degraded because calls will inevitably take longer to set up.

One of the advantages of trunked systems is the efficient use of the spectrum. Because the system is two frequency simplex a single channel pair of frequencies is able to support a conversation between two mobiles or group of mobiles. This advantage can be very easily lost if great care is not taken in the design of the control system. There will always be the situation where two mobiles are not in range of the same base site and therefore require the use of two separate channels. In many cases, particularly with time shared control channel systems it is possible to find a mutually acceptable site for both parties. This can be achieved by the use of diversity voting at each site on the control channel and allocating accordingly. On wide area schemes it is also advisable to limit the use of fixed mobile dispatchers and insist on wire line connection to minimise the unnecessary use of spectrum. In single site systems this is not required as a single frequency pair would support both dispatcher and mobile.

6.7   Other trunking techniques

6.7.1   Transmission trunking

So far in this chapter only message trunking has been discussed. That is, a call is set up on a radio frequency channel which remains allocated for the duration of call. If the signalling speed is increased such that the time taken to send control telegrams becomes negligible then it becomes possible to operate a system with transmission trunking. Each time an operator releases his press to talk (PTT) switch the channel on which he was operating returns to the pool. Both mobiles in a conversation return to the control channel and wait for a new channel assignment. This occurs when the second mobile activate his PTT switch. In this trunking discipline no airtime is wasted due to pauses within a conversation since the channel can be re-assigned immediately to another user. It has been shown that a significant increase in mobiles per channel can be accommodated for the same grade of service [5]. This approach has its dangers in that unacceptable wait times can occur during a conversation in the busy hour as the system approaches its peakload. In order to alleviate this problem a compromise has been used in a practical system [6] and is known as "Quasi-Transmission Trunking". In this case part of the pool of channels is reserved for calls in progress during the peak period. Also a timer is used to determine whether a channel should be returned to the main pool during a conversation. If users are quick in their exchanges then the same channel will be maintained for that call. If not, then a new channel will be assigned. The algorithm involved attempts to maintain an average maximum queuing delay of 1 second. The resultant system shows a practical advantage over message trunking, but with increased complexity and control channel overhead.

6.7.2 Self trunking

As mobiles have been more and more intelligent, then it has become possible for them to find free channels without the help of a system controller. This is very useful when operating in the single frequency simplex mode where no talkthrough station is involved. The Japanese personal radio system (PRS) uses this technique, but more recently in Europe the Digital Short Range Radio (DSRR) system is under consideration by Administrations. DSRR is intended to use some "spin-off" from pan-European cellular in the form of the codec, but will use frequency-division-multiplex access (FDMA) method of channel selection.

In DSSR there are up to 2 control channels and 77 traffic channels. If an operator wishes to make a call then he enters the called mobile (or mobiles in the case of a group call) identity from a keypad. When the send button is activated the radio scans the pool of traffic channels following a particular algorithm. When a free channel is found the mobile immediately returns to the control channel and transmits a selective call message. On receipt of an acknowledgement, it then transmits a go to channel message to the called mobile which acknowledges on the traffic channel.

This technique can never match the grade of service which can be obtained from a centralised intelligence, but has great advantage of ease of frequency allocation. It is proposed that frequencies 933 to 935 MHz be used initially in the UK to try out this technique. The Electronic Engineering Association (EEA) has been instrumental in co-ordinating a group of manufacturers to produce "breadboard" equipment to demonstrate the feasibility. This has been successful and has resulted in a European Specification TR24-04.

6.7.3 Digital trunking

Just as cellular radio is progressing towards the use of digital methods for the transmission of voice signals (GSM), so ETSI and the EEC are promoting the formulation of a European Telecommunication Standard (ETS) for trunking. This task is at present with a technical sub-committee of RES and designated RES6.

From technical submissions already received it is likely that one of two solutions will emerge. The first is the use of narrow band individual FDM channels placed within the existing channel separation raster. The second, which is likely to receive the most favour, is a UK DTI submission for the use of 4 time-division-multiplexed channels on a 25 kHz separated carrier. This improves the use of frequency spectrum by a factor of two over existing 12.5 kHz separated systems without increasing the frequency stability requirements of the radio equipment. This is very important as the first use of digital trunking may be above 900 MHz.

The submission proposes the use of 4.8 kbit/s codec which together with channel coding gives a gross bit rate per channel of 9.6 kbit/s. For four channels the total rate is approximately 40 kbit/s. This high data rate can only be contained within the channel bandwidth by the use of multi-level modulation. The proposal further suggests the use of a form of QPSK as the most suitable modulation technique.

The current time scale for RES6 indicates that such equipment could appear in the market place during 1995.

6.8   References

1.   MPT 1327. 'A signalling standard for trunked private land mobile radio system', DTI, January , 1988

2.   SCHOUTE, P. C. : 'Control of Aloha signalling in a mobile radio trunking system', IEE Conference on Radio Spectrum Conservation Techniques, London, July, 1980

3.   BEAR, D.: 'Principles of telecommunication traffic engineering', IEE Telecommunication Series Book, 1979

4.   MPT 1318. 'Engineering memorandum', 'Trunked systems in the land mobile service', DTI, February, 1986

5.   ZDUNEK, K.: 'Design considerations for trunked radio systems', CH2308-5/86/0000-0194, IEEE, 1986

6.   LUM, E., and ZDUNEK, K.: 'Performance of trunked systems with inter-connect', Energy, Telecommunications and Electronics Conference, Houston, Texas, 1984

Chapter 7

Paging systems

Tony K Sharpe

7.1 Introduction

Paging is a system which attempts to solve the "all times, all places, all people" communications problem. We will see how, with very economic use of the RF spectrum, paging is able to keep very large numbers of people in touch with each other at essentially low cost, with small, light-weight, unobtrusive, bodyworn equipment. To achieve this paging is fundamentally a one-way calling system and relies on the availability of a wire-line or cordless telephone network to support the communication process. From simple beginnings where in-building public address systems were replaced and improved by personal paging systems, very small sophisticated Pagers operating in all frequency bands with powerful message capabilities, and supported by nationwide transmitting and control infrastructures, have been developed.

7.2 On-site paging

In the mid 1950s the problem of calling doctors and nursing staff silently without distressing patients was addressed by the radio technology of the day. The first paging system to be introduced, circa 1957, by the Multitone company, used low frequencies in the range 30 kHz to 50 kHz and was conveniently restricted in range by the use of inductive loop techniques. Each receiver was simply tuned to its own carrier frequency. As more users were required frequency modulation of a 40 kHz carrier was introduced.

Inductive loop systems are still widely used today and are favoured by regulatory authorities because of their excellent spectrum conservation properties. However the cost of installation, extension and coverage prediction makes such systems unattractive in comparison to radio systems. Typical inductive loop systems operate with approximately 1.5A loop current from 25W drivers at 40 kHz and have an inductance of approximately 1.5 mH over 200m to 500m with standard 0.5mm flex. Pagers have typically a 10 µA/M to 30 µA/M sensitivity with a frequency deviation of ± 4 kHz.

On-site paging rapidly expanded during the 1960s and 1970s as frequencies in the 27 MHz to 42 MHz and 470 MHz bands were allocated. Channel separations moved quickly from 25 kHz to 10 kHz in the VHF bands, with amplitude modulation, and remained at 25 kHz in the UHF band, with frequency modulation. Until about 1977 the signalling systems were based on sequential tones in the frequency band 500 Hz to 3 kHz, however today there are many different proprietary digital signalling systems on the market with no compatibility existing between different manufacturers' equipment. Transmitters are typically in the 1W to 5W range, and pager sensitivities are in the range 10 µV/ M to 100 µV/M measured in the best direction; i.e., off the body.

On-site paging is essentially a low capacity system of communication with typically 20 pagers per system and a maximum of 3000 pagers to a system. Speech is often transmitted and digital displays are used. The provision of answer-back facilities with integral transmitters radiating in the 146 MHz to 169 MHz band, 20 mW ERP, from antennas with effective gain of -6 dB, is often a part of an on-site paging system. The use of radiating cables and the integration of intercom, telephone and alarm systems all form part of an installation.

7.3 Wide area paging

On-site paging, with a typical range of less than 2 km, grew as an extension of the internal telephone, public address and intercom system business. However applications for the facilities offered in the on-site situation were soon found for much wider area systems requiring greater range. Firemen's call-out and medical answering services are examples. The early systems of this type required limited address capacity but had excellent sensitivity to achieve maximum range from the relatively simple and, by necessity because of the low capacity, low cost fixed transmission networks. The early wide area paging systems, unlike on-site systems, were seen as an extension of the mobile radio business, and this was reflected in the companies entering the market, the specifications and the frequency ranges.

The concept of offering a paging service more widely than within a private or professional organisation was first introduced in North America in 1963, and then quickly in other countries, notably in Europe and Japan. Table 7.1 shows the significant milestones in the growth of wide area paging. Some of these systems are of considerable importance and will be considered in detail; however, what is not revealed is the multiplicity of incompatible systems that have been and currently are in operation.

Name	Milestones	Date	Location
BELLBOY	City wide public service	1963	Washington
SEMAPHONE	National public service	1964	Netherlands
EUROSIGNAL	Pan-European service	1967	Europe
SWAP	Digital paging service	1969	Canada
MBS	National digital broadcast service	1970	Sweden
SWAP	UK paging service opens	1975	UK
POCSAG	Manufacturers agree common digital standard	1978	London
NTT	Japan opens digital paging service	1978	Tokyo
POCSAG	POCSAG service opens	1981	London
RPC 1	CCIR recommend POCSAG	1982	Geneva
POCSAG	All manufacturers offer POCSAG	1986	World
ERMES	New pan-European standard study	1987	CEPT
POCSAG	Three nation paging service	1990	FR, FRG, UK
ERMES	Pan-European standard	1990	Europe
POCSAG	Pan-European satellite paging	1990	Europe
ERMES	Pan-European service	1992	Europe

Table 7.1 Paging milestones since 1963

7.4 Signalling methods

7.4.1 Sequential tone

The early systems used sequential tone signalling and the later digital signalling. Of the tone signalling systems, a division can be made between those with typically two sequentially transmitted audio frequencies and those with typically five or more. All the tone frequencies used to modulate the RF carrier are in the 100 Hz on 3000 Hz audio band. In order to achieve an acceptable address capacity with only two sequential tones, a high level of FDM is used with a choice made from 40, or in some cases 70, frequencies in that band. Thus, very high Q filters are required in order to identify the narrowly separated frequencies, high stability circuits are required in both transmitter and pager and the RF carrier must be modulated by each tone for a duration related to the Q of the decoding filters. Typically two-tone paging systems have an address capacity of between 1000 and 3500, and a calling rate of between 1 and 0.3 calls per second dependent upon the range of frequencies chosen and the possible use of battery economiser techniques. Sensitivity improvements over other methods of signalling are in the order of 6 dB for two-tone systems because of the small noise bandwidth resulting from the use of high Q filters.

In order to relax some of the critical stability requirements necessary for the two-tone systems and to give increased address capacity coupled with increased calling rate, what is sometimes called 'decimal digital' signalling was introduced. Five or more sequentially transmitted tones are chosen from an effective library of 10 frequencies giving a base 10 system for addressing. In fact, more than 10 frequencies are identified to allow for the tone repeat situation, group call and other additional special functions. While there is, in general, only one preferred series of tone frequencies used for two-tone systems, there are many used for five, or greater, tone systems. The EIA/Motorola five-tone system is based on an arithmetic progression of frequencies with a spacing of 15 Hz allowing constant Q techniques to be used in filter design, and relatively short tone durations, each of 33 mS. This gives a calling rate of 5 calls per second and an address capacity of 100,000 is achievable. Other five-tone paging signalling systems are designed to share the same signalling as mobile radios and provide what is often known as 'overlay' paging. This means that ZVEI, CCIR and EEA signalling systems are employed. A very significant sequential tone system in use in Europe today is Eurosignal. Tone frequencies are within the range 313.3 Hz to 1153.1 Hz based on a geometrical progression. Each tone is transmitted for 100 mS, and calls, each of 6 tones, are separated by 200 mS of the 'highest frequency tone' (1153.1 Hz) called a 'line clear tone'. Thus the calling rate is 1.25 calls per second and the address capacity is with a base ten systems. 1×10^6 addresses but with a possible base of 14, increasing the address capacity to 7.5×10^6. [11,12]. Fig. 7.1 illustrates the Eurosignal structure.

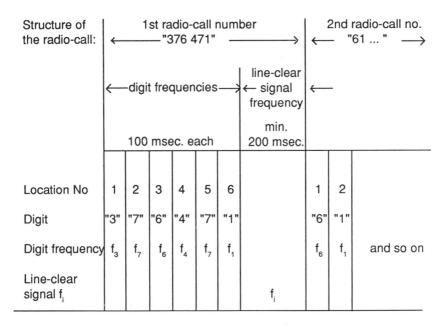

Fig. 7.1 Eurosignal format

7.4.2 Digital

The first truly binary digital paging system was introduced in 1970, in Canada, and subsequently aroused great interest in such methods of signalling. The trend towards digital signalling techniques has not always been fully understood because digital signalling does have some disadvantages. Notably it will in general offer inferior sensitivity, and thus range, because of a greater noise bandwidth than tone signalling. This is especially true if equal or greater address capacities and calling rates are to be achieved. However, as the cost of digital technology falls, and the dual requirement for large address capacity and high call rate rise to meet the requirements of, at first, very large city wide systems and now nationwide systems, digital signalling has come into its own. Digital paging systems also have the advantage that additional functions and features can be easily provided, e.g. multiple pager addresses and text messages. Such features are among the reasons on-site paging systems manufacturers have introduced digital signalling. Before examining some of the more common digital signalling codes, it is important to identify different methods of RF carrier modulation in use. Direct frequency modulation of an RF carrier with an NRZ bit stream has the advantage of a high modulating deviation to modulating frequency (data rate) ratio. The disadvantage is that DC disparity results and thus the modulation spectrum is centred on the low frequencies where, in typical PMR and line systems, significant group delay, attenuation and distortion is encountered. Alternative modulation techniques used in on-site systems, PMR systems and some wide area paging systems, is to shift the modulating spectrum away from DC so that the low frequency signal corruption is overcome. The provision of a sub-carrier or techniques such as Manchester bi-phase coding, on which a binary one is represented by a mark-and-space, and a binary zero is represented by a space-and-mark, achieve this with some success. Such bi-phase FSK systems have the disadvantage of requiring twice the base bandwidth of direct NRZ FSK modulation for the same bit rate. However, all digital systems are wide band and thus subject to distortion. Thus great care in system design is necessary; transmitters specially designed for paging are often required. A summary of the essential differences between the three common signalling methods is given in Table 7.2.

	Sensitivity	Calling rate	Capacity	Message	Cost
Two-tone	Excellent	Poor	Poor	None	Moderate
Five-tone	Good	Moderate	Good	Moderate	Moderate
Digital two-level	Moderate	Good	Excellent	Good	Excellent
Digital M-level	Poor	Excellent	Excellent	Excellent	Good

Table 7.2 Paging systems comparison

7.5 POCSAG

From Table 7.1, it can be seen that the first wide area paging systems appeared in North America and Japan in the early 70s and during the mid seventies digital paging systems were developed by some of the major companies, the Swedish PTT and the NTT. However, during this period one of the most adventurous single channel national paging systems in the world was designed and planned in the United Kingdom by the British Post Office.

System Wide Area Paging (SWAP) was initially proposed in which numerous proprietary digital paging codes could be accommodated by computer controlled paging terminal equipment. Such a system was commissioned for public use in London in 1976. However, it was soon realised that a transmission system combining both simultaneous and sequential techniques was necessary to optimise traffic handling for total country coverage and that because of uneven population distribution a single paging code was also necessary if channel utilisation and occupancy was to be maximised.

The British Post Office realised it was important to select a paging code and a signalling format that would meet performance characteristics they could from experience clearly identify and also at the same time be internationally accepted.

It was with this aim and purpose that in December 1975 the Post Office Code

Standardisation Advisory Group (POCSAG) was established. Under the chairmanship of Mr R Tridgell, 36 engineers representing 13 major paging companies and manufacturing associations met 16 times over a period of two years. Some 68 technical papers were presented and results from research work around the world considered.

The properties of a code ideal for large national or major city use were initially agreed and identified in Table 7.3.

Code capacity	:	10^6 pagers each with 4 addresses
Paging rate	:	6 calls per second
Radio frequency	:	Unrestricted
Radio channels	:	25 kHz/12.5 kHz
Modulation	:	Choice to optimise system performance
Transmission mode	:	Sequential and simultaneous
Adjacent transmitter frequency offset	:	Practical and maintainable at low cost
Line transmission	:	Dedicated with standard modems
Absolute line delay	:	100 μS
Lost call rate	:	2% in limiting typical urban environment or along coverage boundary
False call rate	:	1 in 10^{-8}
Battery economy	:	To be incorporated in code format
Messages	:	Capability for numeric and alpha-numeric containing many characters. A single decimal character was not considered important
Code compatibility	:	Cross calling with other codes to be avoided
Radio practice	:	Not violated

Table 7.3   Ideal code for paging

Finally a code structure and format originally proposed by Philips Research Laboratories (PRL), and in particular Mr P Mabey, was accepted with some minor modifications as a code that satisfied all the requirements identified above, while being non-proprietary and giving to no manufacturer a precedence.

The characteristics of this POCSAG paging code are fully described in the POCSAG report published in 1978 [6] and are summarised in Table 7.4.

The code was initially evaluated by computer analysis at PRL using bit error distribution recorded in London under multipath, shadow and co-channel fading conditions. Finally the British Post Office also completed a comprehensive computer analysis using recorded bit error patterns at VHF obtained from a research project in Birmingham University. The results gave success and falsing rates under many different environmental conditions and in each case these were compared with the same parameters obtained from a known proprietary code. Field trials with a number of POCSAG papers were conducted in London.

Address capacity	:	8×10^6 (2×10^6 pagers)
Paging rate	:	15 calls per second
Message rate numeric	:	5 calls per second (10 digit message)
Message rate alpha	:	1.07 calls per second (40 character message)
Data rate	:	512 b/s (+10 ppm)
Channel separation	:	12.5/25 kHz
Modulation	:	(NRZ) +4.5 kHz (+2.5 kHz) = 0
		− 4.5 kHz (−2.5 kHz) = 0
Transmission	:	Simultaneous or sequential
Lost call rate	:	2%
Error correcting potential	:	2 bits random 5 bit burst per codeword
False call rate	:	10^{-8}
Battery economy	:	18% battery use during normal transmission
Messages	:	Yes (alpha-numeric etc)
Message capacity	:	Unlimited

Table 7.4   The POCSAG code

Sixteen pager addresses are batched into eight unique time slots (frames) every 1.0625 seconds and in this way code capacity, signalling rate and battery economy are all optimised. Because Philips agreed not to apply for patent rights but offered the code for free international use, a high performance world standard paging code has emerged.

The POCSAG code format is based upon 32 bit codewords comprising a (31,21) BCH codeword with an appended overall even parity bit. Sixteen such codewords are concatenated in the form of 8 frames together with a leading synchronisation codeword to form a *batch*. A call is determined by receipt of the synchronisation codeword together with an *address codeword* received in a specified *frame*, which frame being determined from the pager address. Since a pager is only required to scan for its address in a known specified frame, once synchronisation is obtained, battery economy may be maximised by the pager switching off for all but its own frame. Batches may be concatenated ad infinitum, but on commencement of transmission, a preamble, consisting of data reversals is transmitted, to ensure that all pagers utilising battery economy are primed to receive synchronisation, and thence the first address codewords. Message codewords without limit may be transmitted immediately following the address of the destination pager, each message codeword replacing an address codeword within the batched structure.

The POCSAG code is designed for constant bit-rate transmission of 512 bits per second. Fig. 7.2 shows the POCSAG signal format.

Each pager is allocated one of the eight available frames per batch, and one combination of address codeword bits 2-19. Thus 21 bits identify a pager's response code. This is called a Radio Identity Code (RIC) and is normally marked on the pager case. Associated with each RIC are four addresses determined by the two function bits (20/21) transmitted. These address functions have been designated four alerting cadences and in display pagers indicate type of message formatting. A pager can be programmed with more than one RIC. Each RIC can be allocated up to four telephone numbers.

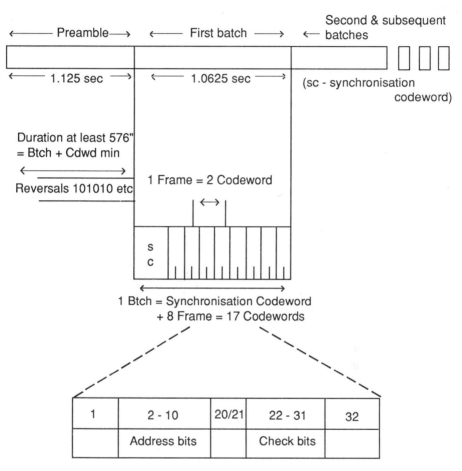

Fig. 7.2 Diagram showing signal format (POCSAG)

In February 1982 the POCSAG code was accepted by CCIR as recommended No. 1 code for international paging. All pager manufacturers now conform to the POCSAG standard and all national paging services offer POCSAG.

The message capability of the POCSAG code identified in the summarised characteristics (Table 7.4), has had a marked impression on the development of paging. The structure of the code allows alpha-numeric messages to be transmitted together with the alert only calls on the same radio frequency. This means that simple numeric only message pagers, advanced alpha-numeric pagers and basic alert only pagers can be controlled from the same paging control equipment and introduced as they become available during the life of any national service. A POCSAG message codeword is like the address, a (32,21) codeword. The first bit is a message flag, '0' identifies an address word and a '1' identifies a message word. The 20 information bits from 2 to 21 contain data and are formatted as 4 bit or 7 bit characters. Message codewords are always associated with the immediately preceding address code and can be concatenated to form any length of message. Messages are not restricted to frames or batches although sync words must be transmitted every 17 codewords in accordance with the POCSAG format of Fig. 7.2. Messages are terminated by an address codeword which could be the address codeword beginning a new message to the same or a different pager. Messages can be of any length. Address function '00' identifies the beginning of a numeric message formatted with five 4 bit hexadecimal characters per codeword. Sixteen characters are defined including 0-9, (), SPARE, U, :, Address function '11' identifies the beginning of an alpha message formatted with $2^{6/7}$ seven bit ASCII characters per codeword. The full 128 ASCII characters are defined providing full upper and lower case capability and printer functions.

The wide interest in message paging was demonstrated by the increased international membership of POCSAG when the group was reconvened in 1980 and 1981 to draft a more detailed specification for message formation. The success of POCSAG must to a great measure be attributed to its chairman Mr R Tridgell who has shown that international standards can be achieved and brought together.

7.5.1 Other codes

Although there are other important digital paging codes, POCSAG is more fully described because it has the widest international acceptance. Three other important codes are compared in Table 7.5. Of these, the Golay Sequential Code (GSC) is supported by the largest manufacturer with a full range of products and is in high volume service, notably in the USA. The NTT code is in high volume service in Japan and also sold by NEC in other countries. The Swedish MBS code is in use in Sweden, but has been seriously considered in other countries because it uses an already existing infra structure of broadcast transmitters [9]. Considerable rivalry has arisen between the GSC code, essentially American in origin, and POCSAG, a British code, resulting in claims and counter claims over their relative merits [7, 15].

Characteristics	MBS	GSC	POCSAG	NTT
Codeword Type	Kasami 26:16	Golay 32:12	BCH 32:21 plus	BCH 31:16
Hamming Distance	3	7	6	7
Codeword Error Detection	Burst errors 10 Random errors 2	Burst errors 11 Random errors 6	Burst errors 11 Random errors 5	Burst errors 15 Random errors 6
Codeword Error Correction	Burst errors 5 Random errors 1	Burst errors 5 Random errors 3	Burst errors 5 Random errors 2	Burst errors 7 Random errors 3
Codeword Efficiency	$\frac{16}{26} = 0.62$	$\frac{12}{23} = 0.52$	$\frac{21}{32} = 0.66$	$\frac{16}{31} = 0.52$
Code Address capacity	1×10^6	409,000	over 8×10^6	65,000
Number of bits per Address	52	60.5	32 (+3 implied) +2 overhead preamble	34.9
System bit rate (bit/s)	1187.6	300/600	512/1200	200
Call rate per sec.	22.84	4.95	15.06/35.3	5.73
Battery economy		No	Yes	Yes
Radio channels in use	FM Broadcast 88 to 104 MHz	150 MHz 450 MHz	470 MHz 150 MHz 930 MHz	250 MHz
In Service since	1978 (Sweden)	1973 (USA)	1981 (UK)	1978 (Japan)

Table 7.5 Paging code comparison (tone only)

The GSC code has two bit rates, 300 bits per second for address codewords and 600 bits per second for message codeword, with differing code structures for each type of codeword: Golay sequential for addresses and interlaced-BCH for message codewords.

The format of the GSC shown in Fig. 7.3 and compared in Table 7.5, is for tone-only address paging without battery economy. The interlacing technique of assembling the words in rows and transmitting in columns is designed to give improved burst error correction capability.

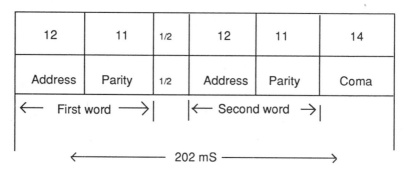

12	11	1/2	12	11	14
Address	Parity	1/2	Address	Parity	Coma

←— First word —→| |←— Second word —→|

←——————— 202 mS ———————→

Fig. 7.3   Golay sequential code address word

The relative differences in the vital parameters of digital signalling code structures, of lost call rate and false call rate, are only marginal and vary according to message length, speed of movement, orientation and environment of the pager. There is always going to be a slight trade-off between these parameters and the system parameters of calling rate, address capacity and battery economy techniques.

7.6 Paging receivers

The fundamental user requirement, irrespective of signalling method and systems infra structure, for a paging receiver is that it should be small, light, unobtrusive, low cost, robust, have a long battery life, and, of course, be reliable under all conditions. To meet all these simultaneously has created some interesting challenges for pager designers. To offer reliable service the RF performance must be compatible with other mobile receivers operating in the same frequency bands. If a pager is to satisfy most of the specifications issued by national PTT organisations [1, 2], then in summary the performance must be:-

Frequency	138-174 MHz
Channels	25 kHz and 12.5 kHz
Modulation	POCSAG - Radio paging code No. 1 (CCIR 584)
Frequency offset	± 1010 Hz
Frequency stability	2 ppm over 1 year
Sensitivity	10μV/m (8 positions average on body on 30m test site)
Selectivity	- 65 dB at ± 25 kHz (or 12.5 kHz)
Spurious response	- 70 dB at ± 50 kHz
Intermodulation	- 50 dB unwanted signals + (25 and 50 kHz)
Desensitisation	- 70 dB at ± 25 kHz
Spurious radiation	2 nanowatts
Battery life	650 hours

The parameter that causes most difficulty is sensitivity, because it is adversely affected as size of pager and power consumption are reduced. As previously stated the signalling system also has a marked effect of sensitivity. For example typical two-tone pagers would be expected to have a sensitivity of 4 μV/m at 150 MHz. The specification for Eurosignal which is a 6 tone system with amplitude modulation on 87 MHz, calls for 2 μV/m in any direction in free space, or on the body 8 position average of 2 μV/m. Sensitivity is also subject to considerable dispute over acceptable and repeatable test methods.

Pagers have inefficient integral antennas and are normally worn on the body, and as might be expected this has a marked effect on antenna performance. Some regulatory authorities and major operators thus specify that measurements must be made with the paging receiver body worn, while others, because of the difficulties in obtaining repeatable measurements, specify free space sensitivity. These latter measurements have the disadvantage that an acceptable sensitivity figure for field coverage calculations is not obtained. The free space measurements can be made in a test cell or stripline while the bodyworn measurements need a 30 metre test site. Such test sites are costly to construct and to calibrate. The recently published IEC standard 489 Part 6 - Selective calling - receiver decoder equipment; RF receiver decoder parameter measurements, described a 30 metre test site together with a sensitivity test method which involves the use of a simulated man, an acrylic column of salt solution and requires an up-down technique of making measurements.

Most pagers today use a frame or loop antenna for frequencies above 100 MHz, although some manufacturers prefer ferrite antennas. The effect of the body on antenna performance is shown in Fig. 7.4. The body removes the two relatively large nulls but does attenuate the RF signal when the maximum body thickness is presented to the transmitting antenna. This effect is especially marked at higher frequencies. Since the weakest field strength conditions are most likely to be found in buildings, it is argued that the field will always have a Rayleigh distribution, due to multipath effects, and that free space and body worn measurements taken in a test site give false figures for calculating the true effective field sensitivity.

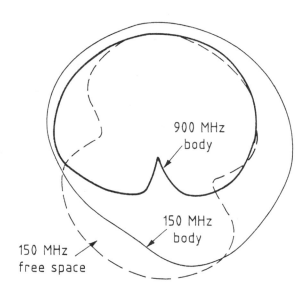

Fig. 7.4   Effect of body on pager antenna performance

Whatever test methods are used the antenna has to have a low Q so that it is not adversely affected by the presence of the body. The RF front end noise figure of the receiver is normally in the order of 4 dB, whilst the antenna gain is only approximately -23 dB relative to a dipole (for a typical 150 MHz paging receiver).

The traditional receiver architectures for pagers are conventional double or single superheterodynes. Low power, one volt circuits have been perfected with typically 10.7 MHz or 21.4 MHz first IF and 455 kHz or 30 kHz second IF. Front end filters have been used and SAW revolutionary direct conversion architectures have been introduced [13, 14]. Such receivers have effectively no IF but directly convert the FSK frequency components of $f_c \pm \Delta f$ into a digital output bit stream. This technique has the advantage that it can be almost fully integrated on a single chip and eliminates costly crystal filtering since most filtering is achieved at audio frequencies.

The digital decoders (or tone decoders) are generally fabricated in CMOS. Some manufacturers use available microprocessors, other custom LSI chips. The techniques employed for obtaining bit synchronisation in digital decoders vary, some manufacturers use a synchronous system in which continuous multisampling of the bit stream finds frame synchronisation without the need for a preamble comprising many data transitions. The POCSAG preamble comprises at least 576 bits of bit reversals to aid bit synchronisation and to contribute towards battery economy techniques. The power requirements, acquisition times and decoder complexity are all related to the bit/frame synchronisation techniques used. It is an area, together with battery economy techniques, over which many patent actions have been fought. The way in which pagers recover from long fades, and the percentage time that a receiver must be powered to guarantee maximum reliability are influenced by decoder performance in terms of data synchronisation as well as the inherent properties of the code format.

The POCSAG code uses batches and frames to give battery economy. The RF circuits which require most of the battery power (typically 5 mW) are switched by the decoder (typical power 200 μW) in accordance with the code format. In the POCSAG case this is only during the synchronisation codewords (32 bits) and the one specified frame period (64 bits). However, the RF circuit and its interface characteristics determine what establishment time must be allowed before an acceptable bit stream can be presented to the decoder. This establishment time can have a major influence on battery life. If the bit rate is increased as for 1200 b/s POCSAG or new ERMES systems the time the RF circuits must be on before good data can be decoded becomes a larger proportion of the batch period unless the batch structure is changed. From 512 to 1200 b/s for POCSAG can increase average power dispatch by 30%. During continuous transmission of codewords a typical POCSAG pager will be dissipating approximately 24% of its maximum receiver power and only 8.5% when receiving no RF carrier or a carrier not modulated with POCSAG codewords. Battery life is thus dependent not only on the pager on time but also the traffic loading of the paging network. Although high currents of 200 mA are used for altering or display illumination, 'on' average duration of seconds per day, dissipates only negligible extra average power.

Bit error correction characteristics of the digital codes are used in different ways. If only one combination of, say 32 bits as in POCSAG is acceptable with perhaps, no more than 2 bit errors, simple correlation techniques can be employed, and even with a few acceptable combinations, many decoders use such correlation techniques on synchronisation and address data. For data message codewords in which one bit random error occurs, correction requires considerable computing power. In addition, the probability of falsely correcting such message data rises dramatically. The Motorola code uses eight interlaced codewords to give an effective increase in bit error correcting capability. However, burst error correction techniques can also be used, with good results on the POCSAG code.

Decoders are also required to store in memory many calls received, each of a different type, and to be able to generate specified and distinctive alerting patterns. A method of programming the individual pager addresses is a further requirement which is achieved by differing techniques, the most popular has been the use of a fusible link diode or transistor array, but today EPROMs are employed.

Message pagers require multi-digit LCD devices and suitable driving circuits. The greater memory requirements of message pagers have resulted in an increase in functions required by the pager user. Radiation from logic circuits, i.e. decoders, micro-controllers, display drivers, into the RF front end via the loop antenna can desensitise the paging receiver. This problem must be carefully considered during design.

Message pagers have normally 16-40 character displays and can store many different messages with memory capacities of 2000 characters. The ability to off load stored message data onto a printer is also available.

7.7   Propagation

The standard techniques for predicting RF field strengths apply equally well to paging systems as to other radio systems. However, because of undefined antenna orientation, the problem of identifying building penetration attenuation and the relatively short duration of each address codeword, safety margins are required and also paging system performance is defined in terms of paging probability as a percentage of potential locations in a coverage area. In any paging system coverage area there will invariably be some deep nulls, for example in lift shafts or underground stations, where the RF signal level is highly attenuated or shadowed. Another propagation hazard is in the construction of buildings with reflective glass windows which not only eliminate ultra violet radiation transmission into the building, but also RF radiation. In such cases, signal penetration may require the use of radiating cables driven from a roof-top antenna.

Many paging locations especially inside buildings are subject to multi path propagation and the RF field strength can then be shown to vary with position according to a Rayleigh distribution. This means that to achieve a 50% probability of successful paging for a static pager the pager sensitivity must equal the median level of RF field strength in the near vicinity of the pager location. To increase this probability of the pager being located in a position where the RF field will be equal to or greater than its sensitivity to say 90% an additional 9 dB in the median level is required.

For much of its life a pager is stationary, however, it is subject to movement when the user is walking or travelling in a vehicle. When moving the receiver will pass through deep fades and the effect of these will be influenced by (a) the rate of movement, (b) length of message, (c) the median level of the RF field, (d) the bit rate of the code, (e) the error correction algorithms employed, (f) frequency of the RF carrier. Because of the trade offs already identified between sensitivity, falsing and information throughput, each code has its supporters and also responds differently in the moving environment. Table 7.6 gives some sensitivity comparisons which are typical for either POCSAG or GSC [15, 7]. On the other hand the interlaced code used in the GSC is claimed to be more sensitive in the long message moving environment and could save transmitters; however due to the lower capacity, more RF channels, and hence transmitters, would be required to serve the same number of subscribers.

Mode	Static	3 mph	30 mph
Tone only	0 dB	1 dB	2 dB
Numeric 10 digit	1 dB	3 dB	3 dB (5)
Alpha 40 character	2 dB	7 dB	5 dB
Alpha 80 character	3 dB	10 dB (17)	7 dB (13)

The figures in brackets are computer simulated results for POCSAG with only 1 bit error correction [7] but were not found in field trials [15].

Table 7.6 Relative median RF signal levels for 90% success rate. 150 MHz

Following extensive testing in London of POCSAG message pager [15], Tridgell and Denman proposed the following equation to calculate the additional field strength required over that needed for tone only paging for 80 character Alpha message paging:

$$\text{Message loss (dB)} = 3.2 + 0.05\varpi$$

where ϖ is the percentage of time spent walking, over the range 0 - 50%.

From an analysis of typical behaviour patterns it is expected that between 15 to 30% of a day is spent walking. The loss of sensitivity at higher speeds is not seen to be a problem, for these speeds will only be achieved in the street and thus not suffer the same loss due to building penetration which must be used to guarantee adequate coverage within buildings, see Table 7.7 below. The penetration loss of a motor vehicle is in the order of 8 dB and the penetration loss within a building will vary with building type and floor.

In the main it can therefore be said that higher field strengths are required for successfully receiving 80 character messages for the common condition of walking inside buildings.

7.7.1 Simultaneous transmission

Wide area paging systems, sometimes called WAPS, by definition must provide a very large coverage area. To achieve this, within the constraints of narrow band mobile radio, many fixed transmitters must be used. The maximum effective radiated power normally permitted in Europe is 100 watts, although in some cases higher powers are allowed e.g. in remote districts or near broadcast frequency band allocations.

Quasi synchronous transmission is used to achieve the required area coverage and the characteristics of some of the paging codes reflect this. For example the POCSAG bit rate originally specified as 512 bits per second which is also its baud rate allows differential delays in the transmission paths to each transmitter modulator of 250μsecs. These are calculated in the following ways by assuming a maximum allowable confusion period [5] of 25% of a signalling element, i.e.:

Paging signalling rate:		512 b/s	1200 b/s
25% of signalling element	=	488 μS	212 μS
10% allowance for telegraphic distortion			
(± 5% isochronous)	=	195 μS	84 μS
TX rise tone tolerance	=	25 μS	25 μS
Radio propagation delay differential			
(10 Km path difference)	=	30 μS	30 μS
Therefore, max. permissible line			
differential delay	=	238 μS	70 μsecs

The choice of offset frequencies for paging transmitters radiating simultaneously has caused some controversy over recent years. The POCSAG report identified offsets of 500 Hz and 2000 Hz because it was influenced by a Japanese paper [5] which shows that offsets should be at least twice the signal fundamental frequency. The POCSAG report was also followed by Telecom Australia and the CCIR Rec 584 on this aspect. Because later work shows that the cancellation and enhancement caused by the two (or more) RF signals adds another Rayleigh distributed field to that which already exists, the offsets have little effect on signalling probability, but do have a marked effect on receiver design problems, particularly as the channel separations are reduced and bit rates are increased. So offsets of ± 50 Hz (max. ± 200 Hz) are normal, and preferred by receiver designers.

The effects of frequency offset, signal level and differential signal delay in a two signal simultaneous laboratory test with POCSAG pager receiving a 40 character message are illustrated in Figs. 7.5 to 7.7. The nearer the two signal levels are to the pager sensitivity threshold and to each other the lower is the message success rate. The greater the offset according to reference [5] the greater the message success whilst the influence is reduced as the signal level increases.

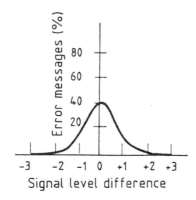

Frequency: 150 MHZ
Offset frequency: 20 Hz
Ref level above threshold 15 dB
Bit rate: 512b/s

Fig. 7.5   Two POCSAG signal simultaneous reception 40 character messages

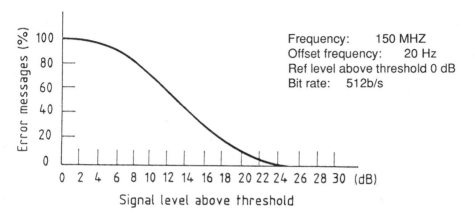

Fig. 7.6   Two POCSAG signal simultaneous reception, 40 character messages

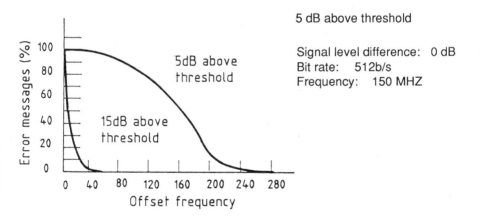

Fig. 7.7   Two POCSAG signal simultaneous reception, 40 character messages

The effect of differential signal delays are illustrated in Fig. 7.8. The ability of the pager to accept very large changes in peak frequency deviation will limit the rate of degradation at large differential signal delays. The waveforms are the effect observed on the output of a frequency discriminator of a receiver in a near equal signal level area in a field trial due to differential signal delays.

Sample period   4 runs   0 hits
Magnification   1.000 ms/div
Magnify above   10.00 µs sample
Cursor moves   730.0 µs x to o   Freq: 149 MHz

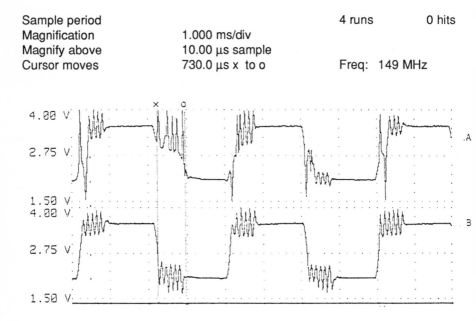

Fig. 7.8   Frequency discriminator output showing effect of differential
            delay between signals from TXA and TXB in actual systems:
            (A) - signal level from TXA slightly greater;
            (B) - signal level from TXB slightly greater

7.7.2   Zone sequencing

To cover a country the size of the UK with 100 watt transmitters about 500 sites
are required. As it is not necessary to radiate all calls in all areas, but, for system
capacity reasons, desirable to radiate different calls simultaneously in different
areas, a mixture of sequential and simultaneous transmission is often employed.
A zone structure is used so that two zones transmitting together do not overlap
but are separated by zones that are transmitting in a different time slot. Major
zones and minor zones can be identified to relate to population densities with
more time allocated to major zones. Fig. 7.9 illustrates a commonly used arrangement.

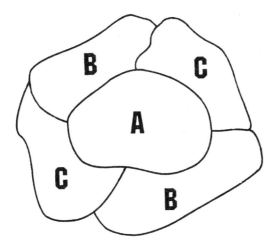

Fig. 7.9   Zone structure for paging

A zone structure as shown does add more complexity to the propagation models used and can result in a need for more sites and thus cost to overcome co-channel interference and re-use distance problems. The Eurosignal system uses a zone structure but, unlike the UK system, separates zones by frequency and not time. This has the advantage of not reducing the calling rate (which, for Eurosignal is very low), but has the disadvantage of increasing the complexity of the pager. There are three zones in Germany, using two frequencies, and seven zones in France, using four frequencies. In Germany approximately 45 transmitter sites are used, radiating at 87.340 MHz and 87.365 MHz, most with approximately 100W ERP, but eight use 2 kW, and two at Hamburg and Stuttgart, radiate 156 kW. All are amplitude modulate except the two 156 kW transmitters which are frequency modulated.

7.7.3   Penetration

The choice of frequency for the ideal paging system is difficult. Propagation losses increase with frequency, but building penetration decreases, as does man made noise.

Low frequencies, below 50 MHz, have co-channel interference problems and channel re-use difficulties. Receiver noise factor increases with frequency and, for given antenna gain, leads to less sensitivity and/or more costly pagers at the higher frequencies. For these reasons where the spectrum has been available the 100 to 200 MHz band has been chosen for paging.

Some building penetration measurements have been made at different frequencies in Japan and the UK. The lower figures are those made in the UK and shown in Table 7.7, though the figures will vary considerably with the type of building and number of windows, etc.

Frequency	150 MHz	250 MHz	400 MHz	800 MHz
Building Penetration	14-22 dB	18 dB	12-18 dB	17 dB

Table 7.7   Building penetration

7.7.4   Broadcast modes

Subsidiary Communications Authorities (SCAs) paging has been permitted in the USA since May 1983 and similar broadcast service type paging is allowed in other countries.   The paging service makes use of sub-channels of an FM broadcast transmitter. In the USA the FFC has changed its rules so that, (a) sub-channel operation is permitted whether or not the main channel is in operation, and (b) the FM broadcast spectrum bandwidth is expanded from 75 kHz to 99 kHz. Sub-carriers of 67 kHz or 92 kHz are used and occupy 15 kHz each. Any of the paging signalling formats can be used.   Although broadcast transmitters operate at high powers, some even at 100 kW with 2000 feet antenna heights, the power in the sub-channels is only a small portion of the total ERP.

The coverage is designed to service radio receivers with better antenna systems than those found on pagers but nevertheless this approach to paging can also be successfully used.   A new wristwatch pager giving a message service and standard time will be available soon using this technique.   A frequency synthesiser is included.   A very low battery-on duty cycle enables a good battery life to be achieved.

7.7.5   Satellite services

British Telecom Mobile Communications (BTMC) plan to introduce in 1990 a satellite paging service [16].   This service will be in the form of an extra zone on the national paging system.   A 1200 b/s POCSAG data stream is used to modulate the satellite uplink carrier of 6 GHz from the earth station. The satellite converts this to a 1.5 GHz down link frequency and transmits this to a coverage area which includes most of Europe.   A vehicle antenna receives the satellite transmission and converting from 1.5 GHz to 150 MHz the signal is inductively coupled to a standard POCSAG pager.   Trials have shown that on motorways and open roads virtually 100% of messages were received.   In central London and other similar cities and under trees in forest regions about 35% of messages were received. This is the result of shadowing due to the attenuation of signals received at a low angle from geostationary satellites.   The target market will primarily be long distance haulage companies and professional travellers.

A satellite message service has been licensed in the USA by the FFC [17]. This has been associated with a location service and the frequencies have been allocated internationally by the ITU; operation is under the nomenclature Radio Determination Satellite Service (RDSS). Geostar have a service in operation in North America with 1000 vehicles (1989) and Locstar intend to serve Europe and North Africa. Although the service is expected to service trucks, boats, railroad vehicles and aircraft, hand-held terminals will soon be available. The mobile receiver frequency is 2.5 GHz. Spread spectrum is used to improve performance. Band width: 16.5 MHz, information rates: 15.6 kb/s, chip rate: 8 Mb/s, modulation: BPSK, PN sequence: Gold Code. Tests have shown approximately 80% message success rate on first transmission. To enhance performance messages are retransmitted after a 3 minute interval.

Yet another satellite message service is also in pre-release operational service IN-MARSAT Standard C. [18] This is designed to support low cost small size terminals, receive frequency: 1.5 GHz, channel spacing: 5 kHz, modulation: BPSK, symbol rate: 1200 symbols per sec. information rate: 600 bits/sec, code: interleaved convolutional.

7.8   Paging transmitters

Transmitters for paging systems are greatly influenced by the system modulation, the power relating and system switching requirements. However, for most digital paging systems, like POCSAG, the low frequency components in the modulation means that the conventional PMR phase modulators are required. Continuous rating and comprehensive monitoring and control are also important essentials. Because of the modulation limitations it has restricted the rapid introduction of low cost POCSAG pagers to the existing tone signally systems market.

The transmitter frequency stability requirements are, for a quasi synchronous system, of the order of $\pm 2 \times 10^8$ (from -10°C to + 50°C) with a drift of $\pm 1 \times 10^{-7}$ per month. The offset frequencies are not considered as important for straight digital alert only paging as they were, but to ease the design and maximise the performance of the many hundreds and thousands of pagers, the carrier stability is important. A high stability at the Tx allows an increased drift in the pager which avoids servicing pagers and also reduces maintenance adjustments to the transmitters.

Two channel transmitters are used by BT, because time division zoning techniques are used; the transmitter can be switched between channels rather than to off, occupying 50% of the time on channel 1 and 50% on channel 2 in the major traffic zones.

7.9 System architecture

In the USA paging is a major business and has grown by essentially because private operators build local city wide systems, which operate to the most part independent of other cities and their systems. Frequencies, signalling systems and number allocations are all different. Two-tone, five-tone, three and four different digital systems, display and voice are all mixed. In smaller European countries the user would like to move from city to city and use one pager. Thus more integrated systems have been conceived from the outset. These systems either have a single frequency allocated for paging in all cities or a multi-channel pager is required. The problems of frequency allocation, and co-operation between independent operators have been inhibiting factors for National Paging in the USA; an allocation of over 40 channels nationally in the 930 MHz band will overcome this problem.

A transmitter network must be established and some system architectures have been conceived around the availability of an already existing broadcasting network - notably the Swedish MBS system which uses a pilot tone on the national FM sound broadcasting service. A new paging service in the USA is also proposed using the sound broadcast transmitters. The Swedish system is national and so the multi-channel pagers used are costly. Although at first this seems an ideal solution to the nationwide transmission problem, there are some drawbacks; the broadcasters are thinking of new ways of using their outer frequency bands on their transmissions notably for sending data on programme content, and the coverage offered is not as complete as it needs to be for paging. This is because the broadcast receiver has a better antenna in general and the listener knows if he is in range because he is listening and the effective power radiated in the narrow band pilot tone carrier is only a fraction of the very high broad band station power.

The other need of multi-channel pagers is, as mentioned before, the weakness of the signalling system (e.g. Eurosignal). Even if a signal frequency is available and the calling rate possible is very large it is still desirable not to send calls out in all regions unnecessarily. This can be achieved by not allowing paging in country areas, thus creating uncovered zones between cities. In a large country like the USA this is acceptable but in a small country like the UK it is not. The compromise is to reduce the effective calling rate by temporarily switching off the transmissions in the large cities (i.e. areas of high traffic density) and for these time slots transmit in the separating country areas. Many European countries have chosen a three phased zone structure as illustrated in Fig. 7.9. Zones type A, could be a big city zone type, and could be given a greater proportion of the cycle time (e.g. 50%). The number of pagers that can be supplied on one radio channel depends on, (a) the traffic density during a peak busy period, (b) the calling rate of the code, (c) how the pager population is distributed between zones in a zoned system, (d) code address capacity, and (e) the acceptable grade of service in terms of blocking probability and transmission delay. Table 7.8 compares some codes and gives maximum pager populations for a busy hour traffic density of 0.1 calls per subscriber with an acceptable grade of service.

System	Address Capacity	Call rate per second (tone only)	Max. pager population per channel for traffic - 0.1 calls/sub/busy hour					
			Tone only		10CH Numeric		40CH Alpha	
			None-Tone	A-Zone	None-Tone	A-Zone	None-Tone	A-Zone
EIA	3.5×10^3	0.21	3.5k	3.5k	-	-	-	-
EIA	1×10^5	4.70	100k	80k	-	-	-	-
Euro signal	7.5×10^6	1.20	40k	20k	-	-	-	-
NTT	4×10^6	5.50	180k	90k	60k	30k	10k	6k
GSC	4×10^6	3.30	110k	50k	50k	26k	18k	9k
POCSAG	8×10^6	15.06	490k	240k	163k	80k	32k	15k
POCSAG	8×10^6	35.30	1.1M	560k	380k	180k	75k	35k
ERMES*	60×10^6	140	4.6M	2.3M	1.5M	700k	300k	140k

* The figures used here are provisional

Table 7.8   Population coverage for a specified messaging for the various paging systems

The use of sequential zone transmissions is good if most users request regional coverage and can accept a possible delay of say 2 minutes before transmission of a page. If the sequence period is reduced then the effective calling rate is reduced by the need to transmit at least 3 preamble batches (1 second each). The ratio of the major A zone time to the minor B and C zone times could be dynamically adjusted to optimise the system performance especially if everyone wants just one major zone (e.g. London). Other systems allow the user to dial selected zones so that calls are only transmitted in zones where the pager is expected to be. This is good for traffic optimisation, and good for the user because he can be called anywhere in the whole country without paying for coverage in every zone in advance for every day of the year.

7.10   Paging terminal concept

A paging terminal is essentially a computing system that interfaces the telephone network to a network of RF transmitters. It has to perform a number of important functions.

a. Sufficient input trunk ports must be provided to offer the acceptable grade of service for the estimated traffic.

b. A file for each subscriber's telephone number. This file must contain a complete register of information relating to the number dialled.

c. Access to each subscriber file for updating. In the event of failure to pay, change of pager due to fault, or re-assignment by system operator, and change of availability (present/not present in system), or transfer instructions by pager user.

d. Temporary registers to allow queuing of calls prior to transmission. This allows sequential zoning to be served and a choice of signalling systems to be permitted. It also accommodates very high peaks in the traffic demand.

e. The storage of voice messages, allows for either pre-recorded announcements common to all subscribers (e.g. "paging call accepted", "number invalid", "pager not present", etc) or user defined. Such voice storage and retrieval by pager users requires large memory capacity but adds considerably to the usefulness. Calling subscribers can be given information on the availability of the pager user with personally recorded voice messages and invited to leave personal messages only recoverable by the pager owners with security codes.

f. The sequencing of transmissions in a zone structure and the encoding of data into a form suitable for modulating the radio transmitters. The encoding of the modulating signals can be achieved in district controllers which are remotely sited and thus economic on line costs and allow the delay equalisation of lines to the transmitters to be preset and simply maintained.

g. The terminal must have a queuing capacity to group calls in registers relating to frame numbers if POCSAG or similar signalling systems are employed or if multi pager types are to be serviced. The formation of message codewords must also be achieved and a temporary memory large enough to handle the traffic provided.

h. Redundant operation of main and standby equipment is normally necessary to provide the level of system reliability required. Alarm monitoring of remote transmitter equipment is also necessary.

i. Address capacity and networking is also important. To enable system growth and optimise on telephone network interfacing the distribution of call processing and transmitter control is a very useful property.

j. Traffic analysis and possibly billing and call charging is often requested. The ability to record calls enables traffic peaks and the number of calls for each subscriber to be identified. The generation of additional meter pulses is another way of charging for pager calls.

Terminals which handle many different types of pager signalling formats are often called SWAP terminals. SWAP is an acronym for System Wide Area Paging. This was first conceived in Canada to overcome the problem of single source supply and was adopted by British Telecom and other operators in the USA and Australia. However, it does not offer a good system concept because of the time wasted by the transmission of many different preambles and the incomplete batching. The use of voice paging is attractive to the user but not the system operator unless he can obtain enough channels to support the user demand or if the charges can be continually increased without objection. Techniques of voice storage and compression are used to aid this process.

7.11 ERMES

The European Radio Message Service (ERMES) is a new pan-European terrestrial paging service planned for the 1990s. This new service was proposed in a CEPT technical committee and is now being specified as a standard by the ETSI, Technical Committee Paging Systems (PS).

It is hoped that the standard can be issued in time for a service to open in 1992. The objectives are

a. A common network across all the countries of Europe to allow European wide roaming.

b. A standard code format to handle high capacity text traffic.

c. A common European paging specification and acceptance for connection approvals.

7.11.1 Traffic

Traffic calculations have been based on 8% of population of Europe (400m by 1995). It has been estimated that 70% of traffic will be text. To accommodate big city busy hour traffic 25 kb/s of information has been estimated. Although the code format has yet to be agreed it is estimated that the gross information rate will be 75 kb/s. The gross bit rate per radio channel is expected to be approximately 6.6 kb/s.

7.11.2 Spectrum

The operating frequencies will be between 169.400 and 169.800 MHz. Frequency division multiplex will be used with channels of 25 kHz. Each pager will have multi-channel capability and be able automatically to scan 16 channels and select the correct frequency assigned to the subscribed system operator in a given location (e.g. country).

7.11.3   Modulation

In order to maximise the gross bit rate within a 25 kHz channel 4 level modulation is proposed. Tests comparing QDPSK with 4 level FSK have been conducted. Initial results in Fig. 7.10 compared a model receiver in (a) a static laboratory condition, (b) a single transmitter field trial environment moving through a Rayleigh field at 30 and 100 km/hr and (c) a two transmitter quasi synch trial with 40 µsec and 1 dB signal differences. These results are not very encouraging and further tests are being conducted.

The problems arising from increased bit rate are that the pager sensitivity will decrease and the susceptibility to signal delay differences greatly increase. For POCSAG systems the typical separation of transmitters is between 20 km and 40 km. Due to propagation effects resulting from topography and both buildings and bodies equal signal strength areas can often be encountered with flight time differences of 30 km x 3.3 µsec/km = 99 µsecs. This means that the transmitters in an ERMES system must be located on a closer grid. One quarter of a 6.6 kb/s bit period is 37 µsec and 25% of a 3.3 k baud signalling element is 74 µsec. Different techniques for signal synchronisation at each transmitter site must be employed with such high bit rates (e.g. off air standards). This is necessary to eliminate the effects of telegraphic distribution outlined in Section 7.1. The paging transmitter specification will be more demanding in terms of rise-time tolerance and adjacent channel power performance. The choice of 4 level FSK for ERMES was partly influenced by the transmitter specification. The adoption of QDPSK would have called for linear power amplification.

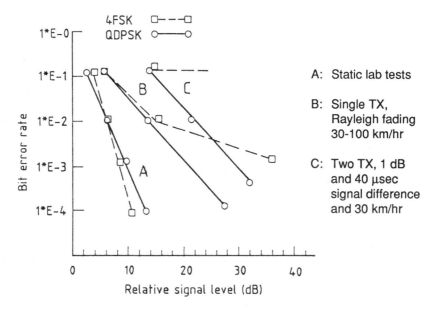

Fig. 7.10   Comparison of 4 level FSK and QDPSK tests for ERMES

7.11.4   Pager performance

The performance of the paging receiver will be affected by the synthesisers in terms of sensitivity, power dissipation (battery life) frequency hopping speed, and adjacent channel performance. It has been estimated that to meet the necessary RF performance, switching times will be of the order of 30 μsecs if the hops are made across only 2 channels and the pager current is limited to 30 mA.

7.11.5   Case structure

The code is awaiting final agreement but issues to be addressed are:

(a)   battery economy techniques
(b)   cycle time (max. delay time of a call)
(c)   address location
(d)   synchronisation system and performance
(e)   country boundary operation
(f)   separation of text and address calls
(g)   narrow cast structure and flexibility of group calling
(h)   repeat message handling

7.11.6   The challenge

It can be seen that there are many interesting challenges ahead for the pager designer, because it must be above all small and low cost. The combination of a pager and a digital cordless telephone (or telecommunication unit) will surely give the most cost effective way of using the spectrum and providing the user with the service needed.

7.12   References

1.   British Telecom specification RC4501D for radiopaging receivers, 31 March 1980, (see also RC4511 and RC 4512 for numeric message display and alpha numeric message display, April 1982)

2.   Telecom Australia.   Specification 1343 issue 2 radio paging receiver (POCSAG digital code) October 1981

3.   IEC Publication 489, Methods of measurement for radio equipment used in the mobile services. Part XX selective calling, receiver decoder equipment, RF receiver decoder parameter measurements , Central Office '75. April 1983

4.   KOMURA, M., HAGIHIRA, T., and OGASAWARA, M.: 'New radio paging system and its propagation characteristics', IEEE Trans. Vehic. Tech. Vol. VT-26.4, Nov. 1987, p. 362

5.   KOMURA, M., et al: 'New radio paging system', Japan Telecom Review, July 1977, Vol. 19, p. 3

6.   British Telecom: 'A standard code for radiopaging', Report of the Post Office Code Standardisation Advisory Group (POCSAG). June 1978 and November 1980

7.   SANDVOS, J.L.: 'A comparison of binary paging codes', IEEE Vehic. Tech. Conference, May 1982

8.   TRIDGELL, R.H.: 'The CCIR radio paging code No. I, a new world standard', IEEE Vehic. Tech. Soc. Conference, May 1982

9.   MAKITALO, O., and FREMIN, G.: 'New system for radio paging over the FM broadcasting network', TELE (English Edition), Vol. XXII, No. 2, 1970

10. MABEY, P.J.; 'Digital signalling for radio paging', IEEE Trans. Vehic. Tech. Vol. VT-30, May 1981

11. CEPT Recommendation T/R6.: Systems Eurosignal

12   SHARPE, K.: 'The Eurosignal radio paging systems', T.C.S. August 1975

13   VANCE, I.A.W.: 'Fully integrated radio paging receiver', IEE Proc. (F), Vol. 129, 1982, Jan, p. 2

14. FRENCH, R.C.: 'A high technology VHF radio paging receiver', IEE 1984 Conf. Mobile Radio Systems, p. 11

15. TRIDGELL, R.H., and DENMAN, D.: 'Experience of CCIR radiopaging code No. 1 (POCSAG) for message display paging', IEE 1984 Conf. Mobile Radio Systems, Sept, p. 194

16. CARPENTER, P.: 'Proceedings of the International Conference on Mobile Satellite Communications', London, July 1989

17. BRISKMAN, R.D.: 'Geostar', Proceedings of the International Conference on Mobile Satellite Communications, London, July 1989

18. The Inmarsat Standard-C communication system, Inmarsat, London, 1988

Chapter 8

Switched networks in support of mobility

Robin Potter and Martin Pigott

8.1 Introduction

Although mobile radio can exist in isolation, for example private mobile radio systems, the increasing emphasis being placed on public mobile radio systems means that most systems need to be interfaced or integrated with the Public Switched Telephone Network (PSTN) or Integrated Services Digital Network (ISDN).

Because of the major investment and huge size of the PSTN, historically mobile networks needed to adapt to the PSTN; however in recent years the demands placed on the PSTN/ISDN by improved and more complex mobile services and networks means that the demands of mobile networks have forced an increasing pace of change on the PSTN. Indeed it is not hard to conceive of the total integration of mobile and fixed networks giving the ubiquity of coverage of the PSTN and the flexibility of mobile networks. This change is likely to be achieved using techniques of the Intelligent Network (IN).

Standards are of key importance in the whole area of networks. The achievement of a telecommunications system that allows world-wide communication is an achievement of standards, in particular the International Telecommunications Union (ITU) and the committee that deals mainly with networks, the International Telegraph and Telephone Consultative Committee (CCITT). Recently there has been a rise in significance of the International Standards Organisation (ISO) in the area of Telecommunications, as the much predicted convergence of telecommunications and computing draws nearer. The influence of ISO is particularly strong in the areas of data and software.

The rise in significance of regional standards-making bodies has also been a significant feature of the last few years, in particular the American National Standards Institute (ANSI) and the European Telecommunications Standards Institute (ETSI) which is formed of the technical committees of the older Council of European Posts and Telecommunications (CEPT). A particular success for ETSI in the last few years has been the development of the GSM recommendations supporting the system as described in Chapter 13.

8.2   Network structure

Generally, all networks consist of some of the following set of elements:

- local exchanges
- trunk exchanges
- gateway exchanges
- a numbering plan
- transmission
- private networks

All these elements are of significance in the support of mobile networks by switched networks.

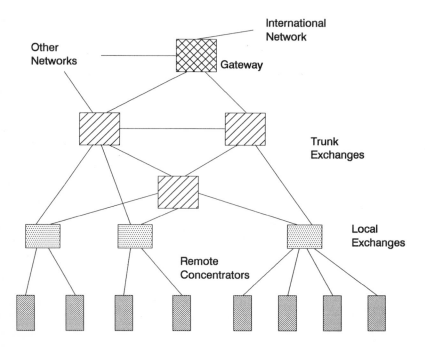

Fig. 8.1 Network structure

8.2.1 Local exchanges

Local exchanges are nodes in the network where traffic is concentrated; the low calling rates generated by customers are concentrated into streams of traffic for onward routing to trunk exchanges. The location of local exchanges is a balance between transmission and switching costs, and the availability of sites. With the rise of digital technology, local exchanges are supplemented by small units often known as Remote Concentrators which are directly connected to the Local Exchange. Digitalisation of networks has allowed the replacement of a number of small local exchanges with a set of remote concentrators.

In modern digital networks the local exchange (or remote concentrator) also provides the 2/4 wire conversion which was previously performed at trunk level in older analogue networks. Numbers of local exchanges are obviously greater than trunk exchanges but again are dependent on the relationship between switch, transmission and accommodation costs. Charging is generally performed at the local exchange with billing information passed to the central billing system either over dedicated links or sometimes over the signalling network.

Traffic from mobile networks is generally not connected to local exchanges as the traffic has already been pre-concentrated in the mobile networks, so connection is more appropriate at trunk level.

8.2.2   Trunk exchanges

Trunk exchanges in general act as tandem switches and operate on traffic pre-concentrated by local exchanges. In older analogue networks trunk switches, e.g. Group Switching Centres (GSCs) switched in 2 wire mode but were usually interconnected by 4 wire transmission links. In addition transit networks were sometimes provided which operated in 4 wire mode and provided guaranteed transmission loss to avoid the necessity of interconnection of all GSCs.

Modern digital trunk exchanges operate in 4 wire mode (all digital systems are 4 wire systems, ISDN access for example provides a 4 wire path over a 2 wire pair by using sophisticated transmission techniques) and are interconnected with digital transmission systems. Customers are occasionally connected directly to trunk switches either to offer special features or for reasons of topology and economy.

Specialised trunk networks may also be employed offering service features not offered on the generality of trunk exchanges or because calling patterns are somewhat different. A good example of specialised networks is the Digital Derived Services Network (DDSN), a part of the British Telecom network that offers 0800 and 0345 services. Specialised networks are used when the traffic volumes make it difficult to justify the expense of upgrading the main trunk network or where service features are offered by a different vendor. However, the cost of switching through the trunk switches and specialised network switches may be high if volumes start to rise significantly. In some cases the service will then be offered on the main network and the specialised network can be further developed or re-engineered to provide other services.

The trunk network is of most significance for interconnect with mobile networks as it is dimensioned and designed to handle pre-concentrated traffic. An additional feature of trunk exchanges are that they are often interconnected using Signalling System No 7 which has some advantages for the mobile radio operator (greater repertoire of signals and improved post-dialling delay etc).

8.2.3   Gateway exchanges

Gateway exchanges perform the function of interworking between networks, either between national and international networks or between dissimilar networks e.g. PSTN/ISDN and Packet Switched Public Data Networks (PSPDNs). Major features of gateway exchanges are usually the ability to perform complex signalling and service interworking, the ability to extract accounting information (used to settle accounts between networks, and not to be confused with charging the customer which is generally a function of the local exchange), and support of a greater range and variety of interfaces (e.g. national and international variants of signalling systems).

8.2.4   Numbering plan

The original numbering plan for the international telephone network is described in CCITT Recommendation E.163. In this scheme each country or zone was assigned a country code of 1, 2 or 3 digits, with a maximum international number length of 12 digits (excluding prefixes such as 0 and 010 etc). This scheme served until the early 1980s when consideration was given to the numbering aspects of the ISDN. It was recognised that the numbering for the ISDN would evolve from the PSTN so the numbering plan for ISDN is an evolution of the existing E.163 Recommendation.

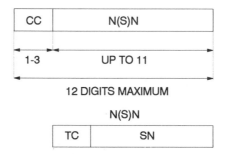

CC: Country Code
N(S)N: National (significant) number
TC: Trunk Code
SN: Subscriber Number

Fig. 8.2   E.163 number structure

Recommendation E.164 "The numbering plan for the ISDN era" was approved in 1984. The main principles of E.163 remain with the introduction of an increased international number length to 15 digits, the introduction of a network destination code (NDC) to embrace the E.163 trunk code and identify special networks. In addition, E.163 permitted analysis of up to 2 digits to determine the international route, this gave a variable maximum number analysis of 3 to 5 digits depending on the length of the country code, whereas in E.164 this was fixed at a maximum of 6 digits including country code. This change reflected the increased capability of Stored Program Control (SPC) exchanges over earlier Strowger or Crossbar exchanges. Additionally a sub-address was defined which could range in length from 4 to 10 digits (or characters).

Recommendation E.165 details the arrangements for the implementation of the E.164 plan and specifies (unusually for CCITT) a date of 31st December 1996 for bringing the Recommendation into effect.

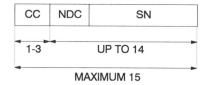

CC: Country Code
NDC: Network destination code
SN: Subscriber number

Fig. 8.3   E.164 number structure

In July 1989, OFTEL published a consultative document on "Numbering for telephony services into the 21st Century". This document made proposals for changes to the numbering scheme in the UK by adding an additional digit to increase capacity and indicate the service called. This scheme would operate within the framework of E.164 and would increase the availability of numbers within the UK. A feature likely to emerge from the new numbering plan is the provision of a Personal Number which can identify an individual and have portability between networks.

The identification plan for mobile subscribers is contained in CCITT Recommendation E.212, and the allocation of mobile station roaming numbers is defined in E.213, shown as Fig. 8.4. This is further referred to in Chapter 13, however, as the PSTN/ISDN operates on E.163/E.164 addresses, the mobile network will be required to pass only those types of addresses on to the PSTN/ISDN.

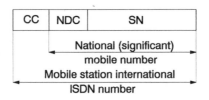

The number consists of:
-Country Code (CC) of the country in which the mobile station
is registered, followed by,
-National (significant) mobile number which consists of National
Destination Code (NDC) and Subscriber Number (SN)

Fig. 8.4   Mobile number structure

8.2.5   Transmission

The last ten years has seen the introduction of digital transmission systems both nationally and internationally, initially on co-axial systems and more recently on optic fibre systems. This has led to reduced costs and increased bandwidth available to customers within increased reliability of transmission systems. In the PSTN/ISDN environment the basic speech/data channel is 64 kbit/s, with 30 channels multiplexed into a 2048 kbit/s (2 Mbit/s) level as the basic input rate to the transmission multiplex equipment.

In the local network between customers and the switches the actual transmission rate over the line transmission systems is either 144 kbit/s (ISDN 2 x 64 kbit/s + 16 kbit/s for signalling) or at 2 Mbit/s.

In the main transmission network between switches, higher order transmission systems are employed at rates of 8.448, 34.368, 139.264 Mbit/s and sometimes higher rates such as 565 Mbit/s. These links are generally provided over optical fibre links, but digital trunk microwave radio systems also provide a significant amount of transmission capacity.

While the 64 kbit/s basic channel for speech (using A-law PCM) or data, forms the basis for modern fixed telecommunications networks, lower rate speech coding is being deployed on expensive international transmission links using 32 kbit/s Adaptive Differential PCM (ADPCM). In private networks, rates below 32 kbit/s are already in use, and mobile networks such as GSM/PCN will use 13 and 6 kbit/s rates over the radio channels.

8.2.6   Private networks

Private networks have grown in importance over the last few years with availability of both digital transmission and digital PBXs. Private networks are usually set up for reasons of security, increased functionality over the PSTN or to handle specialised services e.g. Data.

Within the UK a special signalling system called Digital Private Network Signalling System No 1 (DPNSS) has been co-operatively specified by network operators and vendors. This offers normal PABX like functions over a private network including such services as Call Completion to a Busy Subscriber (CCBS) not currently available on the PSTN. The specification offers the capability for interworking between switches (PABXs) from different vendors.

8.3 Switching systems

The first type of switching system was the human operator, and there is a view that the history of switched networks is merely an attempt to recreate the flexibility and capability of the human operator! The functions of switches can be summarised in general terms as:
- detection of "off hook" calling condition
- initial path set-up to register (or register record with modern SPC systems)
- reception of keyed digits
- call set-up through the exchange
- ringing or alerting the customer
- detection of answer from the customer
- supervision of call during the conversation phase
- detection of release by either customer-removal of connections across the exchange
- recording of billing data

It is worth noting that not all of these functions are performed by all switches.

8.3.1 Strowger exchanges

These now very old switches are based on the mechanical two-motion selector and are still used in many parts of the world. The arrangement of the selectors was very flexible, but became difficult when exchanges were interconnected directly rather than via operators. Although the customer's number as dialled remained unaltered, different sets of digits were used to route the call through intermediate exchanges. These routing digits were provided by register-trans-lators which were originally operated mechanically and were later electronic.

8.3.2 Crosspoint exchanges

The term crosspoint switching covers both crossbar and reed relay technologies. This type of switching was developed to offer improved control flexibility over step-by-step systems. Common control elements are used to operate the switch contacts, store digits, and perform signalling.

Originally the switching matrix was composed of mechanical crossbar switches which required not inconsiderable power to operate them, and to hold the switch once operated. Later some manufacturers developed latching crossbar switches which required power to operate and release, but minimal power while the switch path was held. These latching systems were usually associated with Stored Program Control.

Reed relay systems, which comprise encapsulated precious metal contacts, were also used and gave improvements in reliability over crossbar switches where the contact surfaces were generally open to contamination. In crosspoint systems the cost of the switchblock as a percentage of the total system cost reduced, and as a result, the switchblock was sometimes used to associate common control elements like tone senders/receivers or registers. These types of functions required more sophisticated control than could be offered by mechanical devices, and gave rise to the stored program control concept.

Crosspoint systems can suffer from the problem of blocking, where with increased load, the probability of being unable to reach a certain outlet on the switchblock increases. This problem is almost completely overcome in modern digital switching systems.

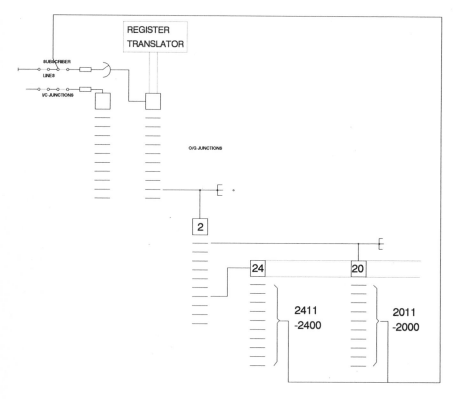

Fig. 8.5 Strowger (step by step) exchange

8.3.3 Stored program control exchanges

The concept of stored programs in exchanges has been in existence for some time, and is not just a product of the microprocessor age, however with the advent of large capacity reliable processors and the microprocessor, Stored Program Control (SPC) has come into its own.

In SPC systems the control functions are carried out on specialised software running on processor(s). As a result of concentration of control into the software, careful design is needed to ensure that failures do not cause the exchange to stop working completely. With the increasing reliability of hardware, which is in many cases duplicated or even triplicated, attention has focussed on producing software of sufficient quality, reliability and performance to give trouble free operation. This is achieved by careful design and extensive testing of software prior to loading onto the control processor and, in operation, recovery from faults is achieved by a number of actions including complete reload of the software as a last resort (which can take several minutes).

One of the advantages of SPC systems is that new facilities and capabilities can be added to the system, without the need for hardware changes. This is especially true for digital SPC systems.

8.3.4   Digital stored program control exchanges

Digital SPC, or as they are now commonly known, digital exchanges combine the flexibility of SPC and the simplicity of digital switching. Digital switches contain no moving parts and as a result are extremely reliable, all switching functions are performed by gates and with improvements in Large Scale Integration (LSI) are becoming smaller and smaller.

Digital exchanges are generally composed of two main elements, Time switches, and Space switches, with the time switch performing switching between timeslots in PCM streams, and space switching performing switching between PCM streams (within the switch). Commonly switches are quoted as T-S-T or Time-Space-Time, where a call passes through the switching stages in that particular order. Other arrangements of switches eg T-S-S-T are also found in practice [1].

Digital switches are ideal for interfacing with PCM systems as the PCM system can be connected directly to the time switch, additionally analogue transmission links can be connected via Primary Multiplexors (PMUX) in bundles of 30 or 31 circuits. Initially many digital exchanges performed all functions within a central processor; however with the increasing availability and power of microprocessors, many functions are performed in a distributed architecture with individual microprocessors performing such function as switch operation, signalling interfacing etc.

With digital SPC systems as the control of the exchange was effected in the central processor, the logical development of common channel signalling systems meant that the two call control processors in a connection could talk together via a common channel signalling link. This avoided the need for complex interfacing from analogue signalling systems, into the digital world of the exchange, and then back out to the analogue network. As a result of this step, the Integrated Digital Network or IDN was created, and the further step to an Integrated Services Digital Network (ISDN) was made possible [2,3].

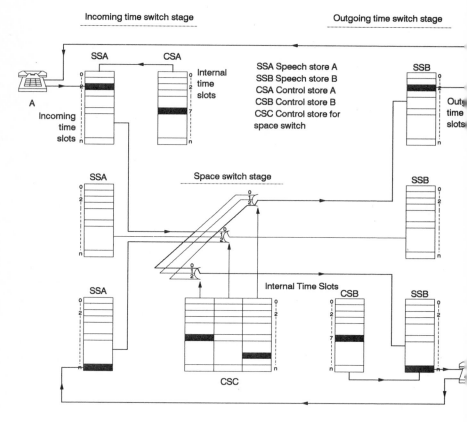

Fig. 8.6 The basic principle of the digital switch

A connection between an A customer and a B customer is set up by loading the control stores concerned with control information in the store positions that correspond to the internal time slot in question. According to this example the A customer speech data is cyclically written in position 2 in the first incoming time switch stage and the B customer speech data are cyclically read out from position n in the third outgoing time switch stage. The internal time slot used in this example is number 7.

Thus the control store of the incoming time switch stage is loaded with address 2 as regards the speech data in the speech store for A and the control store of the outgoing time switch stage is loaded with address n as regards the speech data in the speech store for B.

The control store of the space switch stage is loaded with the address to switch input O, which according to the figure connects together the two time switch stages so that a speech connection is established from A to B. However, a speech path in the opposite direction is also required, since the switch stage provides a 4 wire through connection. The second speech path is set up in the same way by the control stores being loaded with the required control information in another time slot.

When the internal time slot chosen for the connection from A to B (no 7) is connected in, the information in the control stores will address the speech stores and a suitable switch. The A speech data can then be transferred from the incoming time switch stage, via the space switch stage to the outgoing time switch stage. This procedure is repeated cyclically, at a frequency of 8kHz during the whole of the call.

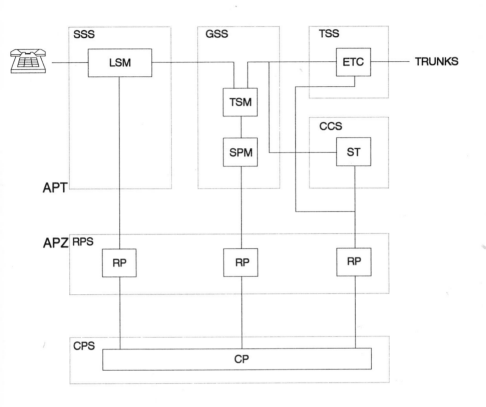

APT Switching system
SSS Subscriber switching system
LSM Line switch module
GGS Group switching subsystems
TSM Time switch modules
SPM Space switch modules
TSS Trunk and signalling subsystem
ETC Exchange terminal circuit

CCS Common channel signalling subsystem
ST   Signalling terminal
APZ Control system
RPS Regional processor subsystem
RP   Regional processor
CPS Central processor sybsystem
CP   Central processor

Fig. 8.7 A typical digital exchange (AXE10)

8.4 Signalling systems

With automation of the telephone network, the need to define signalling systems arose, and the history of signalling systems shows increasing complexity and speed. [4]

The term "protocol" is often used in the context of signalling systems: the term merely means a set of defined rules where actions and responses are known and predictable. In signalling system design complete and rigorous protocol definition is vital if reliable interconnection and operation is to be assured.

A distinction can be drawn between channel associated systems where the signalling information is conveyed within the channel to which it refers, and common channel signalling systems where signalling information is carried within a common control channel (sometimes replicated for security reasons).

8.4.1 DC signalling systems

DC signalling systems rely on DC continuity of copper pairs, with digits being indicated by making and breaking the DC path. The most common protocol used is known as Loop Disconnect (LD). DC systems are mostly used on older exchanges and are slow.

8.4.2 Multi-frequency signalling systems

Multi-Frequency (MF) signalling is now in common use from customer to network and is still in use in some networks (although dying out) for signalling within the network. MF is suitable for use over longer distances than DC systems and is generally quite fast. However it does require accurate encoding and decoding of complex tone sequences and requires careful maintenance.

MF systems usually have two parts, line and register signalling. The line signalling is used to supervise and set up and release calls, and the register signalling is the means to convey digit information. MF signalling systems are usually in-band, although CCITT R2 line signalling is out-of-band at 3825Hz.

Some MF systems are compelled, where the forward tone is held until a positive acknowledgement is received from the far end. Fully compelled systems continue to send the positive acknowledgement until removal of the forward tone is detected.

8.4.3 Common channel signalling systems

With the introduction of SPC it was found more convenient to exchange control information directly between processors using a common data channel. This is the basis of both CCITT Signalling System No 6 and No7. SS No 6 was developed during the late 60s and early 70s and was the first common channel signalling system to see service. SS No 7 was developed from SS No 6 and overcame many of its limitations. SS No 7 is the basis of network signalling in most national and (now increasingly) international networks. [3]

All signalling information is carried within a separate common channel (often Timeslot 16 of a PCM system) which may be routed with the speech circuit it controls (associated signalling) or may be routed differently to the speech circuit it controls (non-associated signalling). The advent of common channel signalling means that there are in effect two networks, the speech circuits and the signalling network.

SS No 7 has a number of component parts: of particular interest for mobile networks are as follows:

Mobile Application Part (MAP) - this supports all the procedures for updating of location registers as described in Chapter 13. It is a highly complex protocol.

Transaction Capabilities (TCAP) - this is a protocol which only contains the rules for decoding messages (MAP in the case of mobiles).

Signalling Connection Control Part (SCCP) - this protocol contains procedures to ensure that messages are routed correctly through the network.

Message Transfer Part (MTP) - this protocol operates at the lowest layer of the protocol stack, and ensures reliable error free transmission in the presence of noise, network failures etc.

Integrated Service Digital Network User Part (ISDN UP) - this carries signalling information for the ISDN and is used to set up and release connections in the ISDN. It will also generally be used in mobile networks for mobile to mobile, and mobile to fixed network calls.

Telephone User Part (TUP) - this is a simple version of the ISDN UP that can carry information for call set up in the telephony network.

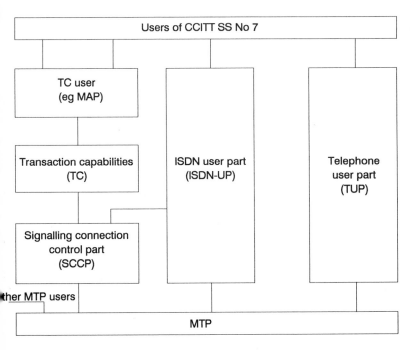

Fig. 8.8 Architecture of SS No 7

(a more complete description is given in CCITT Recommendation Q.700)

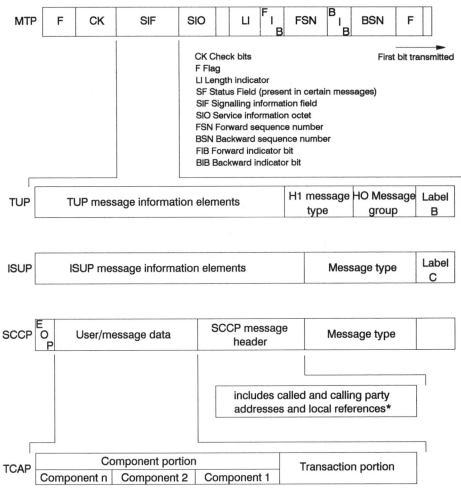

Fig. 8.9   Examples of SS No 7 messages

It is worth noting that MAP, TCAP and SCCP can be used in mobile networks to carry information that is not related to setting up a circuit switched call (such as location updating information) and is a key part of the GSM system.

8.5 Integrated services digital network

The ISDN is a logical evolution of an Integrated Digital Network(IDN), it offers a full range of services to the customer over a single interface, as opposed to earlier networks which may offer a range of services over several types of interfaces [5]. This concept is a major step forward and although it requires that services be presented to the customer over a single interface, it does not require the service to be implemented over a single network. This concept allows an evolutionary path forward from PSTN/dedicated networks to a fully integrated ISDN.

8.5.1 Access

The first problem to be overcome is extending the IDN to the customer's premises. Since the exchange line is 2 wire, two techniques are used to achieve 4 wire operation. Either time division duplex is used (sometimes known as ping-pong transmission) or more recently a form of echo cancellation in which the transmitted signal is subtracted from the combined transmitted and received signal. This technique applies for basic access (64 + 64 kbit/s for traffic [B channels] and 16 kbit/s for signalling [D channels]), whereas for primary rate access (2Mbit/s [30B + 1D]) normal PCM techniques apply.

CCITT has defined interfaces (or more correctly reference points), however these should not be viewed as constraining, and equipment can be designed to cut across the reference points.

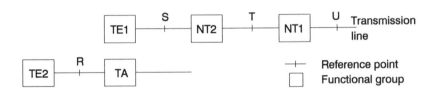

NT1   (Network Termination 1) provides termination of the transmission line.
NT2   (Network Termination 2) distribution to ISDN network access eg LAN or PABX
TE1   (Terminal Equipment 1) equipment requiring the services of the ISDN eg Facsimile, machine etc.
TE2   (Terminal Equipment 2) equipment designed for another non-ISDN interface eg X or V series.
TA   (Terminal Adaptor) used to match non ISDN interface equipment to the ISDN Interface.

Fig. 8.10   ISDN reference points

The access signalling protocol used is known as Digital Subscriber Signalling System No 1 or DSS1 and is contained in Recommendation Q.931. British Telecom specified an access protocol in advance of CCITT, and this is known as Digital Access Signalling System No 2 (DASS2) and has several differences from DSS1.

8.5.2   Network techniques

SS No 7 is used within the main network to support the signalling needed to set up ISDN connections, the protocol generally used is the ISDN UP or internationally in Europe a development of the TUP known as the TUP+ is employed. This is likely to be superseded by the ISDN UP in the course of time.

8.6   Intelligent networks

The concept of an Intelligent Network (IN), evolved from the need to offer services like 0800 over large areas. Without an IN the network operator would need to implement translation tables from an 0800 number to an E.163 address at every node in the network. This is possible, but with the requirement to change these translations frequently, the administrative cost would become burdensome, and there would be a significant delay.

The concept of the IN enabled this translation activity to be performed at relatively few nodes, with an increase in ease and speed of updating. Thus at some node in the network the Service Switching Point (SSP) temporarily halts the progress of the call and refers to a database or Service Control Point (SCP). When the SCP has responded, the SSP completes the set up of the call. Many SSPs are able to access the SCP (usually replicated for security), and updating of translations can be performed on the SCP.

8.6.1   Early IN architecture

In early forms of IN, the SCP and SSP were of proprietary manufacture and did not allow a mixture of vendor equipment, they were however extremely effective and successful. The Digital Derived services Network in the BT network is an example of an early IN. However, for service enhancements, the network operator was dependent on the vendor providing new software for the SCP (and possibly the SSP). Clearly this was an unsatisfactory situation for the network operator who wished to increase the range of service offerings and develop new services cheaply.

The early cellular systems (e.g. TACS, AMPS, NMT) were a form of IN, with mobile location information held on a database which was effectively an SCP [4]. However, again the network operator was dependent on the manufacturer both on cost and software delivery times. The Open or Advanced IN attempts to break this cycle.

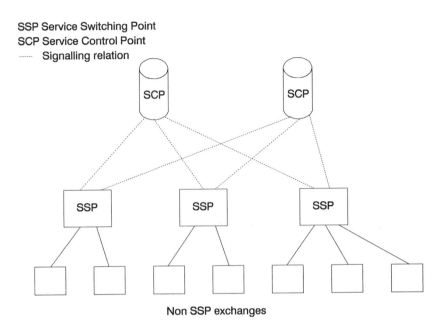

SSP Service Switching Point
SCP Service Control Point
······ Signalling relation

Non SSP exchanges

Fig. 8.11 IN architecture

8.6.2 Advanced IN architectures

Advanced IN architectures will meet a number of key criteria as follows:

- have open interfaces allowing operators to mix equipment from different vendors
- have a service creation environment to allow the operator to create new services without requiring vendor intervention to rebuild the software.
- the ability to customise services to meet customers individual requirements
- offer service as well as network management capabilities
- reduction in the time needed to implement new services in the network

To meet these criteria, a new architecture has been evolved. The situation is still very fluid and terminology is constantly changing, however, the terms SCP, and SSP are still used and in addition the Service Management System (SMS) has been explicitly identified.

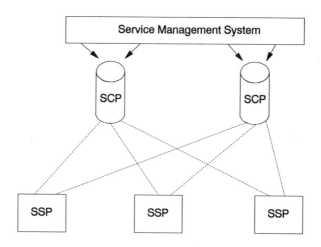

Fig. 8.12   Management of IN provided services by
service management system

In the advanced IN an Application Programming Interface (API) is defined, with the network operator having direct access to this interface in what is known as a Service Creation Environment (SCE). The network operator when he perceives a demand for a new service can enter the SCE and assemble elemental building blocks of a service (often called Functional Components or FCs) across the API and subsequently open the service for public use on the SCP. If the requisite FCs are not available, then dependent on the architecture and commercial agreements, the network operator will either write new FCs or request additional FCs from the vendor. Typical FCs could be "return dial tone", "apply announcement", "analyse digit" etc. It can be seen that all services can be built from these elemental FCs.

Clearly this type of technology presents a technical challenge to the manufacturer, but in addition it also presents a paradox. What incentive does the vendor have to present the network operator with a set of tools which removes the historical hold of vendors over network operators? This advanced architecture may prove difficult to bring to the market precisely because of this paradox, and already vendors are re-aligning themselves with network operators to avoid missing out on this change.

GSM is an interesting example of an IN that has fully open interfaces, but does not currently specify an explicit API.

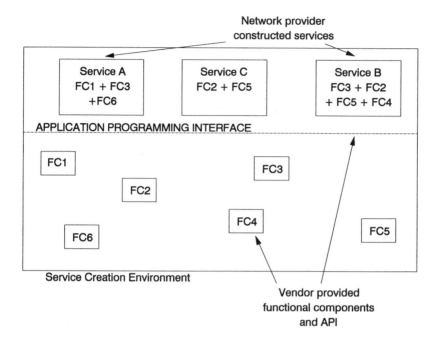

Fig. 8.13 Application programming interface

8.7 Future networks

Networks in the future will need flexibility to respond to a much wider range of demands. More frequent service demand changes can be met by techniques such as IN, however there is increasing demand for bandwidth which can be met by two types of technique.

8.7.1 Broadband networks

In conventional broadband networks, the capability to set up connections of greater than 64 kbit/s is required, typical applications would be video-conferencing, bulk data transfer etc. The demand can be met by circuit switching at multiples of 64 kbit/s or 2Mbit/s, however these type of networks suffer from high costs and generally infrequent use, so the economics can be marginal. What is really required is the capability to offer variable bandwidth dependent on the application. This may be offered in the future by Asynchronous Transfer Mode systems.

8.7.2   Asynchronous transfer mode (ATM)

ATM uses short fixed length packets (or cells) with a minimum of header information to support all types of services. The cells are short and of fixed length to reduce packetisation and queuing delays. A short header is used to allow high throughput switches.

Standards in this area are currently under study in CCITT, although many research and development organisations have working systems. it is likely that these will see service in the next few years, although in the mobile application it is not yet clear where the demand for these high variable bandwidths will come from.

8.8   Network interconnection

The mobile network GSM described in chapters 13 and 14 can provide fully self-contained telecommunications services for mobile-mobile connections. In practice for most telephony/data services, one of the parties involved in the call will be a fixed PSTN/ISDN customer, and interconnection with such networks is a necessity. The resulting structure of such a call path and the network elements thus deployed are illustrated in Fig. 8.14.

Interconnect between mobile and PSTN/ISDN networks will be made using 2 Mbit/s or higher order multiplex structures with 64 kbit/s speech (A-law PCM) or data channels. CCITT SS No 7 signalling will be employed. To enable mobile customers to roam from one Public Land Mobile Network (PLMN) to another, location register updating can be carried out using the Mobile Application Part (MAP) of SS No 7, with transactions to effect this being transported via the PSTN/ISDN signalling transport capabilities. This is an important feature of the European - wide roaming that the GSM standard can offer.

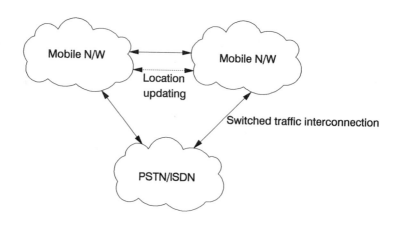

Fig. 8.14   Network interconnection

The advent of all digital networks has eliminated many of the traditional problems of speech transmission levels (dynamic range etc) which were encountered with the interconnection of mobile networks to the PSTN. However one new issue has emerged, modern digital mobile networks generally introduce a significant amount of delay into the speech path with the low bit rate speech coding and the error protecting channel coding that is used to enhance performance/radio spectrum efficiency. The consequence of this delay is to require the use of echo controllers at the interface between the PLMN and the PSTN on connections where voice (as opposed to data) is being used.

8.9 References

1. STALLINGS, W.: 'Data and computer communications' (Macmillan Publishing Co., New York, 1988), Chaps. 7 to 11

2. BREWSTER, R.L. (Ed.): 'Data communication and networks' (IEE Telecommunication Series 22, 1989), Chaps. 7 and 8

3. REDMILL, F.J., and VALDAR, A.R. (Eds.): 'SPC digital telephone exchanges' (IEE Telecommunication Series 21, 1988), Chap. 20

4. POTTER, A.R.: 'Switching and signalling', in HOLBECHE, R.J. (Ed.): 'Land mobile radio systems' (IEE Telecommunication Series 14, 1985)

5. STALLINGS, W.: 'ISDN - an introduction' (Macmillan Publishing Co., New York, 1989)

Chapter 9

The UK cellular system

Malcolm S Appleby

9.1 Background

Public mobile radio telephony has a thirty year history in the UK. The Post Office opened its first service in South Lancashire in 1959, and in 1965 launched its manual System 3 service in London and certain provincial areas. The first automative system, called System 4, opened in 1981 and eventually spread to give near-nationwide coverage.

All of these systems operated in the VHF part of the spectrum, around 160MHz, and were "non-cellular" in design. Their capacity was limited by the amount of frequency spectrum available, so it was clear that any further expansion in mobile telephony would have to be in a different frequency band.

A number of other European countries were meanwhile developing systems for the 460MHz band based on cellular techniques. In particular the four Nordic countries (Norway, Sweden, Denmark and Finland) undertook a joint development of the NMT system, which has gone on to become one of the most successful systems in the world. However the 460MHz frequency band was not available for public radiotelephony in the UK, so no simple solution to the capacity problem existed.

Instead, the UK turned its attention towards a new band at 900MHz which had been allocated to the land mobile service at the World Administrative Radio Conference (WARC) in 1979. The WARC allocation was from 862MHz to 960MHz, and subsequently the Conference of European Posts and Telecommunications administrations (CEPT) decided the details of the usage of the band within Europe. It was decided that the band would consist of 2 x 25MHz segments (one for mobile transmit, one for base station transmit), with 45MHz duplex separation. It was clear that a band of this size and in this frequency range would be ideal for the development of cellular radio.

In the early 1980's, the UK government was keen to liberalise telecommunications and to provide competition. In 1982 it therefore announced that it would be awarding two licences to operate cellular networks. The two networks had to be technically compatible so that customers could elect to change their network subscription without having to change their mobile equipment; both networks were required to open service by March 1985, and to expand to provide coverage to 90% of the population by the end of 1989.

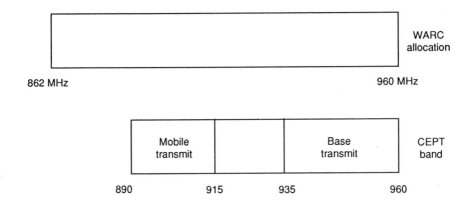

Fig. 9.1   The 900 MHz cellular band -
WARC and CEPT allocation

One of the licences was awarded to British Telecom on condition that it form a joint venture company to run the service as a "hands-off" subsidiary. BT duly formed such a company with Securicor, and Telecom Securicor Cellular Radio Ltd (which trades under the name of Cellnet) was born. For the other licence, a competition was held by the government, and from the five applicants, a consortium known as Racal-Millicom (now known as Racal Vodafone, a member of the Racal Telecom group) was selected.

The standard for the UK networks was decided jointly by Cellnet, Vodafone and the UK government, and was based on the north American Advanced Mobile Phone System (AMPS) standard. The UK standard was called the Total Access Communications System (TACS).

Both Cellnet and Racal Vodafone opened service in January 1985, three months ahead of their licence requirement. Fuelled by competition and a highly buoyant market, both companies expanded their coverage rapidly, reaching the 90% population coverage requirement in 1987, two years ahead of the licence requirement. At the end of 1989 each network was supporting a customer base of around 400,000, and continuing to grow at such a pace that each network approximately doubled its size every year.

9.2   The cellular concept

The basic concepts behind cellular radio are not complex, and indeed are not new, dating back to early work by Bell Labs in 1949, and recorded by them in extensive detail in [1]. The area to be covered by the system is divided up into a number of regular cells, and each cell has a radio base station positioned to give radio coverage of that cell. The total number of radio frequencies, or channels, available to the system is split up into a number (N) of channel groups or sets. The channels are then allocated to the cells, one channel set per cell, on a regular pattern which repeats to fill the number of cells required. Thus each channel set may be re-used many times throughout the coverage area.

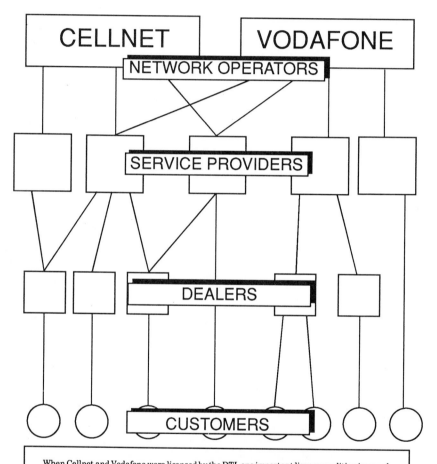

When Cellnet and Vodafone were licensed by the DTI, one important licence condition imposed was that neither was permitted to sell directly to the end customer. Instead all products and services were to be offered via Service Providers who became responsible for setting end-user tariffs, customer contracts, billing, etc. In practice, the 50 - 60 Service Providers operate both directly to customers, and via over 1000 Dealers, who operate at point-of-sale. Thus there is competition at three levels, stimulating the market and driving down prices, and resulting in very rapid growth in the cellular industry.

Fig. 9.2   UK cellular industry structure

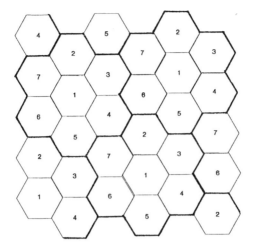

Fig. 9.3 Regular cellular layout (N=7)

9.2.1 Interference vs capacity

When a channel is re-used there is a risk of co-channel interference within a cell from a base station in another cell on the same frequency. However, provided the distance between co-channel cells (D) is kept large enough in relation to the radius of the cell (R), i.e. the so-called D/R ratio is high enough, the probability of interference can be contained to controllable levels.

Fig. 9.4 Frequency re-use - D/R ratio

Clearly, as the number of channel sets (N) is increased, the number of channels per cell reduces, and therefore the system's capacity falls. However as N is increased, the distance between co-channel cells (D) also increases, so the risk of interference reduces. Selecting the optimum number of channel sets is therefore a compromise between quality and capacity. Note that only certain values of N lead to regular repeat patterns without gaps; these are 3, 4, 7, 9, 12 (and then multiples thereof).

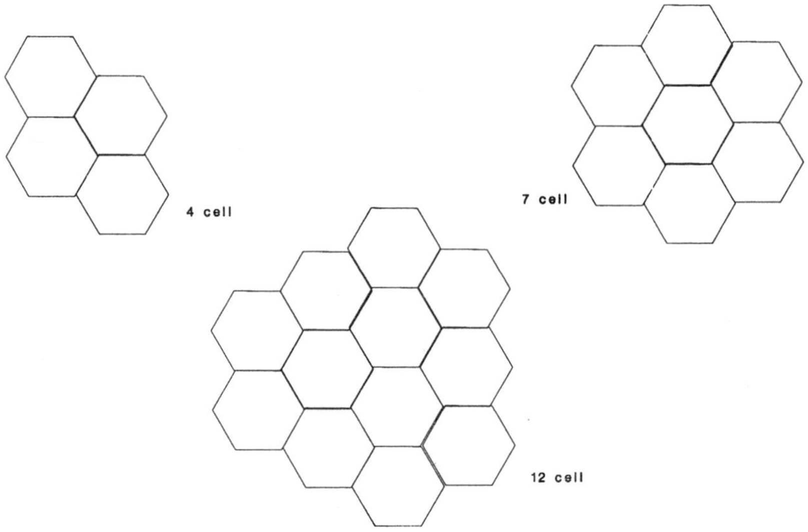

4 cell

7 cell

12 cell

Fig. 9.5   Commonly used cellular repeat patterns

9.2.2   Cell splitting

For a given value of N, the capacity of a system may be increased by reducing the size of the cells so that the total number of channels available per unit area is increased. In practice this is achieved by the process of "cell splitting", where new base stations are established at specific points in the cellular pattern, reducing the cell size by a factor of 3 or 4. By repeatedly splitting cells, the cell size, and thence the system capacity, can be tailored to meet the traffic capacity require-ments demanded by customer behaviour in all areas, from low traffic rural areas, where the cells may be 20km radius or more, to high traffic central urban areas, where cells may be as small as 1km radius. In practice, the variation in propagation, particularly in urban centres, and the accuracy in position to which base stations can be located are the main factors limiting the minimum cell size.

9.2.3   Sectorisation

As cell sizes are reduced, the propagation laws in force indicate that the levels of co-channel interference tend to increase. In a regular cellular layout, co-channel interference will be received from six surrounding cells all using the same channel set. Therefore one way of cutting significantly the level of interference is to use several directional antennas at the base stations, with each antenna illuminating a sector of the cell, and with a separate channel set allocated to each sector. There are two commonly used methods of sectorisation, using three 120 sectors or six 60 sectors, both of which reduce the number of prime interference sources to one. The three sector case is generally used with a seven-cell repeat pattern, giving an overall requirement for 21 channel sets. The improved co-channel rejection in the six sector case however, particularly the rejection of secondary interferers, results in a four-cell repeat pattern being possible, giving an overall requirement for 24 channel sets.

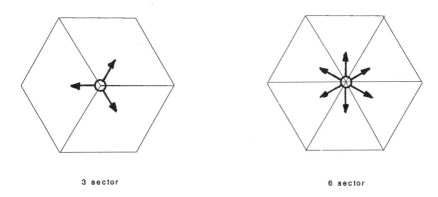

3 sector   6 sector

Fig. 9.6   Sectorisation of cells

A disadvantage of sectorisation is that the larger number of channel sets required results in fewer channels per sector, and thus a reduction in trunking efficiency. This means that the total traffic which can be carried for a given level of blocking is reduced. However the ability to use much smaller cells through sectorisation greatly outweighs such drawbacks, and the end result is a much higher capacity system.

9.3   The cellular "system"

The basic concepts of frequency planning and re-use, and the control of co-channel interference are equally applicable to PMR systems and to traditional non-cellular radiophone systems. Indeed the concepts also apply to tv and radio broadcasting. What is different with cellular is that the individual base stations are knitted together in a way to form a complete system, offering continuous coverage with minimum user inconvenience. There are two key features of cellular systems which make this possible, mobile location and in-call handoff.

9.3.1 Mobile location

When an incoming call is received for a mobile station, the call has to be routed to the cell where the mobile is located so that the call can be connected. One way of finding the mobile would be to transmit a calling message (page) for the mobile on every cell site in the network. However, with several hundred cells and hundreds of thousands of mobiles, the signalling capacity required would clearly be too high. Instead the cellular network is split up into a number of location areas each with its own area identity number. This number is then transmitted regularly from all base stations in the area as part of the system's control information. A mobile station, when not engaged in a call, will lock onto the control channel of the nearest base station, and as it moves about the network, will from time to time select a new base station to lock to. The mobile station will check the area identity number transmitted by the base station, and when it detects a change, indicating the mobile has moved to a new location area, it will automatically inform the network of its new location by means of a signalling interchange with the base station. In this way the network can keep a record of the current location area of each mobile, and need therefore call the mobile only within that area.

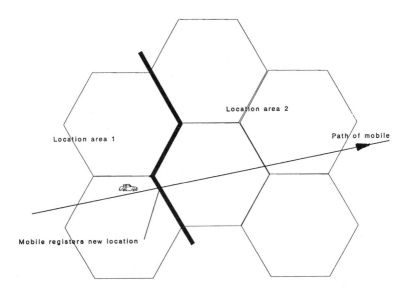

Fig. 9.7   Mobile location registration

9.3.2   In call handover

When a mobile station is engaged in a call, it will frequently move out of the coverage area of the base station it is communicating with, and unless the call is passed on to another cell, it will be lost. The system continuously monitors the signals received from mobiles engaged in calls, checking on signal strength and quality. When the signal falls below a preset threshold the system will check whether any other base station can receive the mobile at better strength, and if this is the case, the system will allocate a channel for the call on the new base station, and the mobile will be commanded by a signalling message to switch to the new frequency. The whole process of measurement, channel allocation and handover takes a few seconds to complete, but the user will only notice a break in conversation of 200-300ms as the handover itself is carried out.

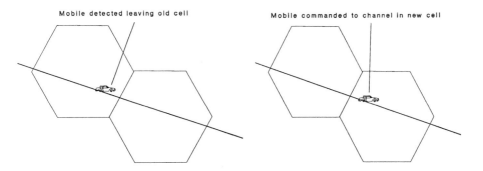

Fig. 9.8 In call handover

9.3.3 Handover and power control in cell planning

Effective and reliable handover is not only highly desirable from the users point of view, but essential in the control of co-channel interference and thereby the maintenance of the cell plan, particularly as the cell size is reduced. A mobile operating on a non-optimum cell will, in effect, be operating outside the cell designated for that area. In other words, the cell boundary will have been "dragged" beyond its planned limit, and this will inevitably give rise to levels of co-channel interference above that planned for the system.

A further weapon in the fight to control co-channel interference is that of mobile power control. So long as the base station is receiving a signal of adequate strength from a mobile, there is no need for the mobile to be transmitting at full power, so the base station can command the mobile to reduce power by sending a signalling message. Clearly by reducing a mobile's power, the likelihood of its causing interference can also be reduced, thus helping to control interference levels.

9.4 The TACS standard

The UK TACS standard [2] was derived from the North American AMPS standard [3], and differs in a number of respects. These differences were necessary since the exact frequency band and channel spacing were different between the USA and the UK, and because AMPS was not immediately suitable for a nationwide two-operator environment. The opportunity was also taken to include some extra features into the standard, and to include a specification for the audio characteristics of the mobile station which was consistent with UK telephony standards [4].

The basic system parameters are as follows:

Frequency band:

890 - 905 MHz mobile tx }	TACS 600
935 - 950 MHz base tx }	channels
890 - 915 MHz mobile tx }	TACS 1000
935 - 960 MHz base tx }	channels
872 - 905 MHz mobile tx }	E-TACS 1320
917 - 950 MHz base tx }	channels

Channel spacing:
    25 kHz (with 12.5 kHz offset)

Voice modulation:
    FM, 9.5 kHz peak deviation

Signalling:
    8 kbit/s with Manchester encoding

Signalling modulation:
    direct FSK, 6.4 kHz deviation

Signalling protection:
    5/11 fold redundancy plus BCH block code

9.4.1 Frequency band

Originally TACS was specified to cover the complete 2 x 25 MHz mobile band (i.e. 890-915 MHz and 935-960 MHz). As an option, mobile manufacturers were allowed to make 600 channel sets which covered 2 x 15 MHz (i.e. 890-905 MHz and 935-950 MHz) since initially the UK government indicated that this was the only part of the band which was to be immediately allocated to cellular. Subsequently the government, in alignment with other European administrations, formally announced that the top 2 x 10 MHz (905 - 915 MHz and 950-960 MHz) were to be reserved for the pan-European digital cellular system (GSM), and would not be allocated to the TACS network. However, some spectrum below the CEPT cellular band was made available for TACS in certain geographical areas. This part of the spectrum has become known as the "extended TACS", or ETACS, band.

9.4.2 Voice modulation

TACS uses a relatively high peak frequency deviation for voice modulation of 9.5 kHz. By comparison, traditional private mobile radio (PMR) systems with 25 kHz channel spacing have a peak deviation of 5 kHz. This wider deviation, together with the wider filters required in the receiver to pass the signal without distortion, results in a degraded adjacent channel rejection compared with PMR systems. However the benefit is that co-channel rejection is much improved. Since adjacent channel interference can be minimised by careful frequency planning, this improvement in co-channel rejection is very worthwhile, and allows systems with a smaller cellular repeat pattern (and hence higher capacity) to be engineered.

9.4.3 Signalling

For the majority of its control interchanges, TACS uses high speed digital signalling with high redundancy to give high reliability even in poor radio conditions. Each signalling word is repeated five times, and at the receive end a bit-by-bit majority decision is performed which corrects the vast majority of burst and random errors. The signalling word is further protected by a BCH block code which is used to check for residual errors, and allows for up to one remaining bit error to be corrected.

On the base to mobile signalling link on voice channels, where it is particularly important that handover commands are correctly received, often under poor and rapidly degrading radio conditions, signalling messages carry an even higher redundancy, being repeated eleven times.

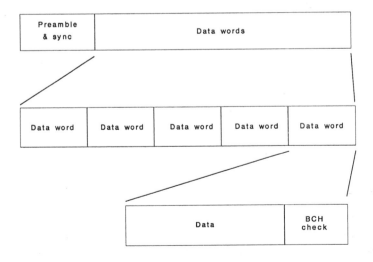

Fig. 9.9 Signalling message format on control channels

Some signalling states are indicated not by discreet digital messages, but by a so-called signalling tone of 8kHz transmitted at 6.4kHz deviation. Signalling tone is effectively the output of the Manchester encoder with the input jammed at one logic level, and is used to signal the mobile answer condition, and mobile cleardown.

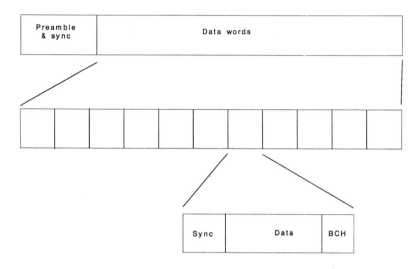

Fig. 9.10   Signalling message format on voice channels

9.4.4   Supervision

When a call is in progress, the base station transmits a continuous supervisory audio tone (SAT) in the frequency range of 6kHz at low level. Three different frequencies of SAT are used by the system, and these are allocated to groups of base stations as an "overlay" to the radio frequency repeat pattern. During call setup, the mobile is informed which SAT is in use by the base station, and during the call it checks whether the correct SAT is being received. If it is not (indicating that there may be a high level of interference), the audio path is muted so that other conversations are not heard, and if the condition persists, the call is aborted. The mobile re-transmits the received SAT to the base station so that a similar supervision can be performed on the mobile to base link.

9.4.5   System organisation and control

Each TACS network uses a set of 21 control channels which carry only signalling. The channel numbers of the control channels are pre-designated, and all mobile stations operating on the network have the numbers permanently programmed in their memory.

Functionally there are three types of control channel:

> Dedicated control channels
> Paging channels
> Access channels

Physically, however, all three functions may be combined on each of the control channels, and in practice this is normally the case. The TACS standard does allow for the separation of paging and access functions into separate physical channels, and for the extension of the total number of control channels in cases where the amount of signalling activity is greater than can be handled by the combined control channels. Since this represents an inefficient use of the available frequencies, it is avoided wherever possible.

The functions carried by the various control channels are as follows:

> Dedicated control channels - unidirectional (base to mobile) - used to carry basic system parameters and information on the channel numbers of the paging channels currently in use.

> Paging channels - unidirectional (base to mobile) - used to carry basic system parameters, location area identity, and paging messages to specific mobiles for incoming call setup.

> Access channels - bidirectional - in the mobile to base direction, used to pass all information to the base station concerning the call setup request or location registration by the mobile; in the base to mobile direction, used to confirm location registrations and to assign a voice channel following a call setup request.

9.5 The cellular network

In essence, all cellular networks have a similar structure, being complete telephone networks in their own right with dedicated exchanges in an interconnected network, and with base stations connected to the exchanges. There are, however, many ways of planning a cellular network in practice, the optimum arrangement for any particular application being dependent upon the capacity required, cost of implementation, capabilities of the chosen manufacturer's equipment etc. As an illustration of one solution to the network design question, the following describes the Cellnet network configuration.

9.5.1 Base stations

For the majority of its network, Cellnet's base stations are organised in a 12 cell repeat pattern with omnidirectorial coverage from each base station. Most base stations have between 20 and 30 voice channels, with one signalling channel carrying all paging and access functions. The signalling transmitter and receiver is fully redundant with two sides operating in worker/standby mode. Voice channels are non-redundant, and any faulty channels are taken out of service, slightly worsening the grade of service of the cell.

The voice channel and signalling channel transmitter outputs are combined using high-Q cavity resonators and fed to colinear antennas of 9db gain. The maximum effective radiated power (erp) of each channel is 100 watts, but in many cases a lower power level is used as dictated by co-channel interference requirements of the overall radio plan.

In the receive direction, most base stations are fitted with six directorial antennas consisting of a colinear mounted in front of a vertical corner reflector. These antennas have 17db gain, and are mounted at 60° intervals round the compass, starting at due north. A preselector/preamplifier is connected to each antenna and the six outputs connected to a switching matrix which allows any one of the voice channel receivers to use any pair of antennas at any time. A pair of antennas is connected to the voice channel receiver so that diversity can be used to minimise the effects of fading on the received signal.

In order that the best antenna is always connected to each voice channel, the base station has a scanning receiver which monitors the signal level on every channel used by the base station via every antenna every few seconds. The results of these measurements are then used by the base station controller to operate the antenna switching matrix. The scanning receiver also carries out measurements of signal strength for both hand-in and hand-out from the cell and for mobile power control. Like the signalling transceiver, both scanning receiver and base station controller are fully redundant.

Using directorial antennas for receiving, even when the base station has a nominal omnidirectorial coverage, carries a number of advantages. The high antenna gain and the use of diversity improves receive performance, particularly for handportables, compensating for the power difference between base station and mobile. Co-channel interference in the mobile to base station direction is reduced since the base station is only "looking one way", and therefore seeing fewer interferers. Also handover processing is improved since the system can establish the direction of the mobile from the present base station, and therefore tie down more closely the "target" cells for handover.

In some urban areas, particularly London, where customer numbers and usage demand a very high capacity service, Cellnet's base stations are arranged in a four cell repeat pattern with six sectors per cell. Each of the six sectors in a four cell cluster is allocated its own set of voice channels. Control channels, however, are allocated on a one per base station basis, with one signalling channel transceiver in the base station operating as a shared resource by the six sectors. When a mobile initiates a call, the base station controller establishes in which sector's coverage area the mobile is located, and allocates a voice channel on that sector. During a call, if the mobile moves to a different sector, the base station controller can within a few seconds carry out a sector-sector handover to ensure the mobile is always being covered by the best sector. In this way, by the effective use of sector-sector handover, cell-cell handover, and mobile power control, Cellnet's base stations manage the radio resource in the most efficient manner, minimising co-channel interference and maximising traffic carrying capacity.

9.5.2 Mobile switching centres

Cellnet's switching network consists of over 20 Mobile Switching Centres (MSC) in 10 locations. The MSCs are digital exchanges with a distributed control architecture, especially adapted for efficient operation in the cellular environment.

Base stations are connected to the switching centres by digital (2 Mbit/s) leased lines. The switching centres are also linked together with digital circuits forming a fully interconnected network. The signalling between base stations and switches, and between switches is proprietary in nature, and is carried in timeslots on the digital circuits.

The MSCs connect to the BT PSTN at a large number of points in order to distribute the traffic load and to minimise the impact of any failures on call handling. Digital interconnection to the PSTN using CCITT No 7 signalling is now used exclusively, having completely replaced the earlier digital/analogue interconnect with loop-disconnect signalling.

9.5.3 Services

The Cellnet network offers fully automatic calling to and from telephones in the fixed network throughout the world, and also provides access to many of the services available on the fixed network, such as information services. In addition, Cellnet's MSCs support a range of "vertical" services to complement the basic cellular service, as follows:

Call divert - all incoming calls diverted to the specified number.

Busy divert - incoming calls to a busy mobile diverted to the specified number.

No answer divert - incoming calls to a non-active (i.e. switched off) mobile, or to an unanswered mobile diverted to the specified number.

Three party conference calling - a third person can be brought into an existing conversation.

Call waiting - an incoming call to a busy mobile is indicated by a tone to the subscriber who can then pick up the second call, placing the first call on hold.

Call barring - selective call barring can be invoked to prevent, for example, unauthorised international calling.

Other services provided by the network include the Cellnet Messaging Service, a voice messaging service which is fully integrated with the cellular network and mobile numbering scheme, and Private Wire, which allows customers to take advantage of lower call charges by linking their private network directly to a

Cellnet MSC. The network supports data services using error correcting voice band modems, with access to BT's packet switching network and IBM's information network via a Celldata inteface. The network also has its own operator service, providing customers with directory assistance and call connect facilities.

9.5.4 System growth

Cellnet's TACS network has grown rapidly since its opening in January 1985. Within a year the network had grown to 100 base stations with over 800 radio channels, and covered London and other major urban centres, and a number of connecting motorways. Table 9.1 indicates the growth of the network since then.

Cellnet began sectorisation in 1986, and by progressive sectorisation, cell splitting and fully utilising the ETACS frequencies wherever available has continued to keep pace with capacity demands. Further expansion is planned for 1990, 1991 and beyond, by which time Cellnet's planned GSM network will be available yielding even higher potential capacity, plus the benefits of digital transmission and pan-European operation.

Date	Number of cells	Number of channels	% population covered
June '86	162	1540	70
June '87	287	4310	90
June '88	354	7040	94
June '89	455	10440	96
Mar '90	560	18000	97 +

Table 9.1   Cellnet Network Growth

9.6   References

1.   Special issue on 'Advanced mobile phone service', Bell Syst. Tech. J., January 1979

2.   'United Kingdom total access communications system mobile station - land station compatibility specification' (Currently available from Racal-Vodafone Ltd)

3.   'Cellular system mobile station - land station compatibility specifications'. EIA interim standard IS-3-B, July 1984 (published by the Electronic Industries Association)

4.   HUGHES, C.J., and APPLEBY, M.S.: 'Definition of a cellular mobile radio system', IEE Proc., 1985, Vol. 132, Pt. F, No. 5, August, p. 416

Chapter 10

The Telepoint system

Professor John Gardiner

10.1 The Telepoint service

The foregoing chapters, 5 to 9, have followed developments of mobile radio systems in two senses - evolution in technical sophistication on the one hand, and growing public access on the other. In the case of UK cellular radio, both developments converge in the Cellnet and Vodafone networks which provide public telephony services (and to a certain extent data facilities as well) supported by an infrastructure which substantially surpasses any closed-user-group PMR system in complexity. The targets in cellular systems proper are to provide a continuous and widely available telephony service to a large community of mobile and portable subscriber units, whether stationary or on the move, and a system with the capability for fixed and mobile network subscribers to contact mobile terminals wherever they may be within the cellular coverage area. The paging service, on the other hand, targets a restricted subset of the cellular characteristics by providing universal coverage to a very large user community of public subscribers but one-way only basic data messages. Plainly there are many other ways of combining various attributes of cellular systems by accepting restricted features in some respects with enhanced features in others. One such combination might address user requirements for full duplex telephony but in a quasi-static operating mode, having the following attributes:

a)   Ready access to the PSTN, even in dense user environments, without saturation problems.

b)   Low cost hand-portable units and call charges comparable with those of the fixed network and appealing to domestic as well as business users.

c)   Public call box type facilities which involve the user in seeking an access locality to initiate a call but without having to rely on fixed pay phones.

d)   Commonality of function between domestic cellular telephone capability and public call box service possibly extending to wireless PABX type operation in addition.

The implications of these attributes may be summarised in turn as follows:

a) suggests a low power system permitting a higher level of frequency re-use than that obtainable from conventional cellular systems. Possibly the use of higher frequencies than the 900 MHz TACS band would be appropriate.

b) points towards a system with a minimal infrastructure cost. The operating cost/infrastructure relationship arises because of the need to support two functions in true cellular radio systems already alluded to, i.e. the ability to track the location of mobiles wherever they are and route calls from the network to them, and to automatically transfer calls from cell to cell in handover. This necessitates a complete mobility management network of fixed links and switching centres to handle the location, registration and handover processes independent of the fixed PSTN/ISDN which ultimately handles the call traffic. The costs of this network impact directly on the pricing regime and subscription charges for users, recognising the upper limit to capacity of the system. In addition, the added fundamental complexity required in the mobiles and portables to interact with the mobility management network inevitably sets a lower limit to terminal (particularly hand-portable) costs.

c) indicates a system in which contiguous cellular coverage might not be appropriate (although not necessarily excluded) and also implies that the capability to sustain handover from cell to cell and to accommodate Doppler shift in received rf signals, might also not be necessary.

d) focuses attention on system aspects where a certain minimum level of functionality in the air interface might be required to accommodate the range of services expected by users.

10.2 The role of CT-2 Telepoint

Plainly, if operating costs are a direct function of the complexity of the mobility management network, then an obvious way to achieve economies is to investigate systems which do not attempt to provide either call handover, or to maintain continuous location data about every subscriber. If these functions are taken away from the full cellular systems then what remains is the capability to initiate calls by establishing access to the PSDN/ISDN via suitable base stations, but not to receive calls from it. Mobility management has now largely disappeared, being reduced to the function of identifying a caller's terminal by its unique number so that call charges associated with both the fixed network and the radio segment can be passed on to the call originator. The service now available is, of course, a reduced one relative to that provided by full cellular networks but experience has shown that users of cell phones initiate many more calls than they receive and, in any case, a system which permits a user to make calls but not to have to deal with unwelcome callers when unprotected by a secretary has much to recommend it even for business users!

The consequences of pursuing such a strategy are, however, that low cost hand-portable terminals become possible and network charges can reduce to a level only modestly greater than those associated with fixed terminal use. The further consequence is also that the latent demand from the general public may prove to be very considerable so that the capability to handle a very large user population, potentially several times that of the cellular user community, must be built into the system from the outset and this dictates operation at very low rf levels in effectively very small cells. In other words, something akin to a generalised cordless telephone type of system in which a handset can access base units anywhere as a connection point to the PSTN, not just at the home base. This was the starting point for the development of the Telepoint service.

It is important to clarify at this stage that "Telepoint" is a term which defines a specific telecommunications service without putting constraints on the choice of air interface used by base stations and portables.

10.3 CT-2 manufacturers and Telepoint operators

The potential of Telepoint-like services had, naturally, become apparent throughout the mobile communications community during the mid-80's. Manufacturing companies embarked on equipment developments, organisations with aspirations towards operating licences began laying plans for networks, but initially there was little attempt at coordinating technical innovation around one general standard in the air-interface. Consequently, a number of different approaches were followed with encouraging measures of success in response to authorisation of UK CT-2 in 1985. When bids were subsequently invited for licences to operate Telepoint services, a number of solutions to the hand-portable/base station air-interface were offered. Such was the perceived urgency to initiate services and meet demand that it was decided to set in motion a stage-by-stage implementation of Telepoint. At the outset, operators would be permitted to install equipment to "proprietary standards" but would be required to undertake to work towards a common air-interface, being set a timescale for implementation of this "CAI" by 31 December 1990.

A further development was also set down as a condition of licence, namely that operators would by mid-1991 implement complete roaming so that once the air-interface standard was in place, a subscriber to any individual operator's network of base stations should be able to access the PSTN by any other operator's bases with charging and billing strategies defined among the operators so that users are billed for use of the system from one source (the operator they subscribe to) regardless of whose base stations handle their calls.

In order to operate a highly competitive service provision environment, four consortia were licenced to provide telepoint services; in three cases these involved interests outside UK. The consortia and their network names are as follows:

Ferranti Creditphone Limited   "Zone Phone"

Philips (Netherlands), Barclays Bank
& Shell oil (USA)   (BYPS) "Rabbit"

STC, British Telecom, France Telecom
& Nynex (USA)   "Phonepoint"

Motorola (USA), Shaye Communications
& Mercury Communications   "Callpoint"

The principal players in CT-2 equipment manufacture emerged as GPT Limited, the originally joint GEC-Plessey Company, Shaye Communications, STC and Ferranti.

10.4 The choice of technology

In looking at the possibilities for a low cost service it might be tempting to target available technology as a basis and in cordless services, analogue techniques are well established world-wide - in Japan particularly where the personal radio service has attracted a large user population since its appearance some years ago. However, in the case of CT-2, although implementation strategies differed significantly, the decisions were unanimously in favour of digital technology. The reasons for this broadly fall under the following headings:

a)   Digital techniques are being implemented increasingly in mobile communication systems, digital signal processing power is becoming available at cost levels which are likely to continue to fall in the short and medium term and any development which stands a reasonable chance of acceptance as a European standard, albeit an interim one, will have to be digital.

b)   Digital transmission in the air-interface offers the advantages already alluded to of robust performance in co-channel interference environments and this advantage is valuable even in situations where contiguous cell coverage is not needed.

c)   Security of voice channels against eavesdropping is becoming a matter of considerable concern and digital encryption techniques are vastly superior to analogue alternatives.

d)   Digital techniques offer the possibility of achieving full duplex operation on single frequency channel but utilising burst mode transmission in each direction at higher than the bit rate required to sustain real time speech in one direction. Whilst in absolute terms this technique does not offer much improvement in spectrum occupancy over two-frequency allocations, its advantage is that duplex services can be introduced in a single spectrum slot, i.e. without the need to look for a paired allocation opportunity. From the standpoint of spectrum managers faced with increasing demands for spectrum, this feature is extremely appealing.

As noted previously, the new Telepoint services have been inaugurated around proprietary standards, but in view of the relatively short time remaining before introduction of the common air interface, the remaining technical sections of this paper will concentrate on the CAI as it has now been defined. It should be noted, however, that no timetable has been set for withdrawal of services based on the proprietary standards and these are expected to be supported by the Telepoint operators for some time after the CAI is up and running, in order to obtain a return on initial investment in fixed equipment and give early subscribers a guarantee of sufficient working life to justify the handset purchase.

10.5 Spectrum and air-interface parameters

In common with all high density systems which rely on the fixed network infrastructure to carry all their voice and data traffic, the complete air-interface specification must address layers 1, 2 and 3 in the OSI nomenclature to establish proper network interfacing and service structure. However, the parameters of the physical layer present a convenient starting point and the question of spectrum allocation is a primary factor.

10.5.1   Spectrum allocation

It should be noted at this stage that the CAI specification arrived at by mid-1989 and now defined in MPT1375, Reference 1, draws extensively on earlier established specifications dating back to 1981. These are listed in References 2-4 and relate to both MPT and British Standards documentation. Necessary extracts from References 2-4, however, appear in MPT1375.

The band allocated to the service extends from 864 MHz to 868 MHz approximately, the channel 1 carrier being assigned to 864.15 MHz and the highest frequency carrier, channel 40, at 868.05 MHz. A tolerance on carrier frequency of ±10 kHz is allowed for at both fixed and portable terminals, and the nominal separation between carriers is set at 100 kHz to support single channel per carrier, time division duplex (TDD) operation.

10.5.2   Modulation and transmitter power

In the interests of keeping handset costs as low as possible, the modulation characteristics have not been set too rigidly. The modulation format is basically binary frequency shift keying with logic 1 taking the higher of the two frequencies as indicated in Fig. 10.1, the peak deviation lying between ±14.4 kHz and ±25.2 kHz. The modulation is shaped by an approximately Gaussian filter, as specified in Reference 2, and as the bit rate is 72 kbits/sec and the bit transitions are constrained to be phase continuous, the result is effectively Gaussian frequency shift keying (GFSK).

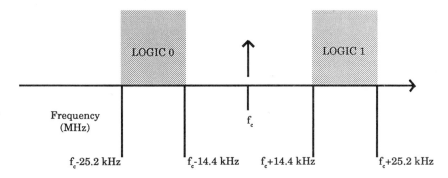

Fig. 10.1   Modulation: Gaussian frequency shift keying
72 kbits/sec data rate

The transmitters in the handsets are limited to a maximum output power of 10mW although manufacturers may opt for a lower power down to a minimum of 1mW. Provision is made for a further low power setting which the handset (or cordless portable part, CPP) can operate on under instructions from the base (or cordless fixed part, CFP). The low power setting is 16 dB ± 4 dB below the normal output.

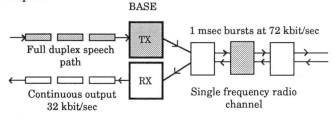

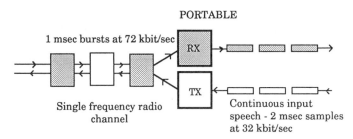

Fig. 10.2 "Ping-Pong" time division duplex

Fig. 10.2 illustrates the basic operation of the ping-pong time division duplex process indicating the basic procedure by which input speech in both directions is sampled and coded (at 32 kbits/sec) and the samples transmitted at the higher data rate which permits time compression of each sample into 1 msec packets. Again there is a choice available to manufacturers to offer either 1 kbit/sec or 2 kbit/sec signalling and this is achieved by permitting two types of multiplex in the traffic channel, MUX 1.2 which operates with 66 bit packets and MUX 1.4 which has 68 bit packets as shown in Fig. 10.3. In both cases, the packets repeat every 144 bits, the interval between transmissions being available for reception. The data packet has a ramp-up and ramp-down time as shown in Fig. 10.4 and allowance is made for up to 1 bit of propagation delay. The suffix element allows for group delay dispersion in the channel filter and the ramp-up and down process is designed to minimise the spectrum dispersion associated with abrupt amplitude changes at the beginning and end of each burst.

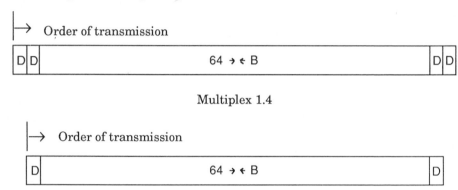

Fig. 10.3 Multiplex 1

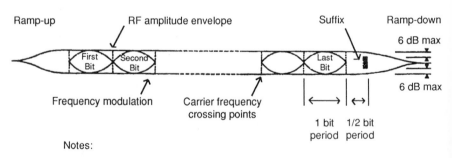

Fig. 10.4 Data packet within RF envelope

10.6 Channels and multiplexs

10.6.1 Multiplex outlines

In common with any telephony system which relies on dynamic channel selection, the CT-2 terminals must exchange signalling information to set up the voice channels before the principal traffic can be carried. However, in time division digital operation, additional activity takes place which relates to bit and burst synchronisation. Layer 1 signalling in CT-2 is responsible for channel selection and link initiation and to do this the digital traffic in each TDD frame is divided among 3 channels: the D channel for signalling, the B channel for voice/data traffic and the SYN channel for burst synchronisation.

The way in which the bits in each frame are shared among the channels depends on the function taking place at the time. Once a link between the CFP and the CPP has been established, traffic is exchanged in either MUX 1.2 or MUX 1.4 corresponding to the frame structure of Fig. 10.3. When a link is being established and the CFP and the CPP are obtaining synchronisation, a different pattern of bits is required in MUX 2 which contains no B channel component, the frame being committed entirely to synchronisation and channel marker data. Fig. 10.5. Clearly timing is all important in the TDD system and the CFP always ultimately takes responsibility for timing - this means that when a portable (CPP) initiates a call, the base station (CFP) must reinitiate call establishment from the fixed end in order to impose its timing on the subsequent link traffic framing. It is this activity, the CFP setting up the link to the CPP, which is the function of MUX 2.

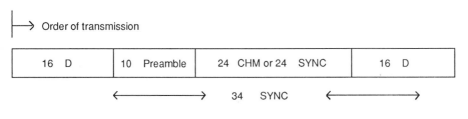

Note: The 66 bit burst is repeated every 144 bits

Fig. 10.5 Multiplex 2

As indicated at the outset, however, since the Telepoint concept does not embrace terminal registration and tracking capability, all calls must begin by the portable demanding access to a base station.

This process requires a third multiplex, MUX 3, which takes account of two considerations:

a)   the CPP on initiating the call has not acquired any bit or burst synchronisation with base station activity.

b)   because the CFPs and CPPs operate in time division duplex, the base station receivers cannot detect incoming signals from portables requesting access while the base transmitters are active.

To make sure that appropriate access requests are received by the base stations during the 1 msec or so per 144 bits when the base transmitters are silent, the CPP repeats a sequence of pre-amble bits followed by synchronisation words for a series of 5 complete 144 bit frames followed immediately by 2 frame intervals during which the CPP listens for a response from the CFP. This exception to the TDD frame structure of MUXs 1 and 2 is necessary so that the base station receiver is guaranteed an opportunity to detect the call request from the CPP and respond. This is MUX 3 and is illustrated in Fig. 10.6.

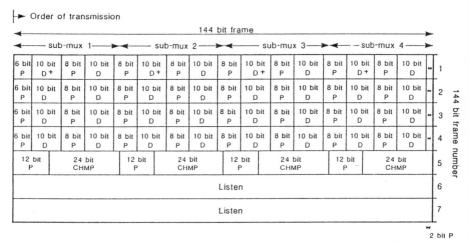

Fig. 10.6 Multiplex 3

10.6.2   Channel scanning and signalling patterns

In order to follow through the sequence of signals and actions which happen during call set up, two further aspects need to be appreciated - the channel selection process and the signalling patterns.

Telepoint has no centralised control of channel allocation for CPPs and CFPs, the cordless telephone apparatus (CTA) must decide what channels are useable whenever a call is to be initiated and use whatever free spectrum they can find. This implies a channel scanning regime to detect both available channels and signals indicating a call request.

Channel scanning and assessment of channel status can be done in either of two ways - by intermittent scanning, pausing on a channel and sampling it in 200 msec intervals for a minimum of 5 samples, or monitoring continuously on a channel for 2 secs. A free channel is defined as:

a)   any channel with a local field strength below an absolute maximum of 40 dB relative to 1 microvolt per metre

b)   where all channels are above 40 dB relative to 1 microvolt per metre then any channel which has the lowest field strength of all those channels available to the CTA service as measured, by intermittent or continuous monitoring, to a resolution of 6 dB.

10.6.3   Signalling patterns

Turning to signalling patterns, these exist mainly in sequences of 24 bits in D and SYN channels. In MUX 1 the D channel data rate averages at 1 kbit/sec in MUX 1.2 and 2 kbit/sec in MUX 1.4, the latter demanding a higher level of performance than MUX 1.2 and naturally, being only useable if both the communicating CFP and CPP can support the higher rate. In MUX 2, the effective data rates are 16.0 kbit/sec in the D channel and 17 kbit/sec in the SYN channel. These are used during the establishment of links between CFPs and CPPs and carry bit patterns which are of two types, Channel Markers (CHM) and Synchronism Markers (SYNC). Preambles of one-zero reversals precede these patterns and provide for clock extraction in the receiver so that the subsequent pattern can be interpreted. The components of MUX 2 and MUX 3 in the D channel comprise two 16 bit and two 10 bit elements respectively in each frame, the division into two segments in each case being implemented to prevent bit sequences in the D channel accidentally reproducing 24 bit CHM or SYNC words in MUX 2.

Both CFPs and CPPs transmit CHM and SYNC words but these differ according to whether they are originated by the fixed or portable ends of the link. Specifically, the CHM pattern sent by the CFP (CHMF) is the bitwise inverse of the CHM pattern sent by the CPP (CHMP). Likewise the SYNCP and SYNCF are bitwise inverses, the objective being to prevent base stations calling each other and any portable directly calling another portable.

The CHM patterns serve the dual functions of marking transmissions in which a CTA is attempting to initialise a radio link and marking a particular time within the multiplex. The CHM patterns contain address information and when a CHMF is detected by a CPP in the vicinity of the base each will check to see if the link establishment attempt is intended for it. Once the CPP being addressed has confirmed that it has received the CHMF, further synchronisation and handshaking need not involve any other CTAs so the CFP and CPP continue their dialogue using SYNF and SYNP patterns which are ignored by other terminals. Fig. 10.7 shows the calling sychronisation patterns in relation to the MUX 2 bit patterns.

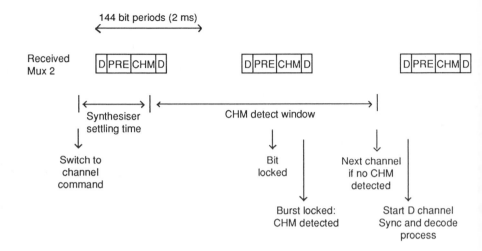

Fig. 10.7 Calling channel detection and burst sync at the CPP

10.6.4   Link establishment procedure

With the aid of the foregoing, it is now possible to follow through the sequence of events which take place during call set up. The portable starts the process by deciding on a useable channel and transmitting an access request in MUX 3. This comprises the bit pattern combinations of Fig. 10.6 with the CHMP elements containing appropriate messages, namely a link request and the portable (CPP) identity code (PID) which is unique to the CPP concerned and is used as the basis for call charging and billing. The access request also contains a signalling *request* bit (SRr) to obtain the maximum signalling rate from the CFP which the CPP can handle. The CPP repeats the transmission continuously until it obtains a response from the CFP, detected during the listening frames or until the process is timed out after 5 seconds.

The CFP detects the MUX 3 CHMP during its scanning cycle, checks the validity of the ID and sets a signalling rate capability bit (SRc) in response to the SRr from the CPP. The CFP then transmits a LINK GRANT codeword in MUX 2 containing a link identity code (LID) for the CPP to use in future communications. When the CPP detects the LINK GRANT word it transmits a handshake codeword looping back the received LID which the CFP then compares with its own transmitted version. If the two do not match then the CFP reverts to channel scanning, but if they do then the link is set up and the CFP and CPP then exchange protocol initialisation messages and corresponding *confirmation*. At this point, a layer 3 message initiates a CFP signalling element in MUX 2 which instructs the CPP to switch to MUX 1.2 or 1.4 according to the SRc status, and both ends of the link are then ready to handle speech or data traffic. It should be noted that neither MUX 1 carries any sychronisation information so that if framelock is lost by some interruption or interference on the radio path then the CPP must go back to MUX 3 and start again. A number of points of detail are omitted in this summary of the set-up procedure, readers are referred to reference 1 for further information and for details of other signalling activities associated with functions such as call termination and so on.

10.7 Speech and channel coding

10.7.1 Speech coding

The coder chosen for CT-2 is the 32 kbit/sec adaptive differential pulse code modulation (ADPCM) standard used in fixed terminal telephony and specified by CCITT. (Recommendation G721, 1988). At first sight this might seem an odd choice in view of the earlier points made about system capacity and the amount of effort that has been put into developing the 13 kbit/sec codec specified for GSM which uses a regular pulse excitation algorithm with a long term predictor (RPE-LTP).

In fact, two considerations ultimately outweighed the benefits of lower bit rate coders - availability and processing delay. The ADPCM codec is already established and available in low cost chips, the speech quality is high and the coding is robust to radio path variations. Processing delay is a major issue because CT-2 remains essentially a cordless extension to the fixed network and therefore must conform to line system standards which only permit a round trip delay of 5 msec in the speech path. The TDD transmission scheme introduces a 1 msec delay, currently only the ADPCM codec can keep the speech processing element within the remaining 4 msecs permitted.

The issue of compatibility with line rather than cellular radio standards is important because of the fundamental difference between the Telepoint service and true cellular radio. Telepoint does not have the mobility management network which is the essential infrastructure for cellular service and which effectively stands between the radio segment and the fixed public network. The CT-2 interface with the line network needs to be as transparent as possible and compatibility of the speech coding across the radio/line boundary is a valuable contributor to this.

10.7.2 Channel coding

A recommended standard for digital transmission in mobile radio systems has been in use since 1981 and is detailed in MPT 1317, Reference 4. This specifies a 48/64 BCH code which requires transmission of data packets which are 64 bits long, and comprise 48 data bits, 15 cyclic redundancy check (CRC) bits and a final bit to provide an even bit parity check on the whole 64 bit word. This format is used for CT-2 signalling messages and, with its capability to detect up to 5 bit errors, provides a good compromise between ease of implementation and error detection performance. Greater sophistication would be hard to justify since, as with any mobile radio transmission system, burst errors are likely to dominate and forward error correction would require extensive bit interleaving among many data packets with corresponding delay implications. A straight forward ARQ strategy is therefore recommended.

10.8 Implementation

The novel features of the CT-2 air interface introduce a number of new aspects to the problem of developing national networks. The most striking of these relate to the way in which the four Telepoint operators will all use the complete frequency band which has been allocated to the service and to the constraints resulting from TDD operation. The low powers radiated by both base stations and hand portables mean that very small cells will be a feature and although all the operators offer nationwide services, this will not necessarily be in contiguous cells in general. Additionally there is no provision in the air interface for timing adjustments to cope with long propagation paths, nor for channel equalisation of the kind developed for the GSM system, nor for accommodating Doppler shifts from rapidly moving terminals. These simplifying features, however, do make for straight forward hardware and all the operators are confident of a large user population which in turn raises the question of system capacity. Estimates based on a cell radius of 63m and a frequency re-use distance of 200m yield a figure of 5000 users/km² and this figure could be greatly enlarged in 'hotspots' of peak demand by deploying smaller cells with vertical frequency re-use within buildings as well as horizontal re-use in the open.

As was pointed out in the opening section of this chapter, Telepoint is a service concept rather than a physical system designed around a specific air interface. It was also noted that one desirable feature of enhanced cordless telephone systems would be the capability to operate in all three common environments - domestic cordless, Telepoint and office wireless PBX. But the demands of these operating situations are very different and alternatives to the CT-2 common air interface standard have been much debated and are being developed, starting from the wireless PBX requirement rather than the telepoint service needs.

For the high density user environment, capacity issues are likely to dominate the air interface design together with other features related to numbers and distribution of base stations within buildings. In the case of the GSM cellular radio system, TDMA has been adopted as the basic access strategy but implemented in a two-frequency assignment scheme. Also, the GSM signal structure contains substantial overheads in channel coding, channel equalisation, propagation delay and Doppler shift correction and so on, which are important functions in a large cell system which is developed to cater for fast moving and relatively high-power mobile terminals. As a result, a basic throughput of 8 x 13 kbits/s requires a total of 271 kbits/s transmitted on the up and down link carriers separately. In a short range system, on the other hand, much of this overhead can be dispensed with since time dispersion of the signal is likely to be very small as will be propagation delay and, in the quasi static operation envisaged in the Telepoint service, so will be Doppler shift.

The merits of time division duplex to achieve duplex operation in a single frequency band have already been commented on and there is the possibility, as higher and higher bit rates become possible in the air interface, of combining TDMA with TDD.

This approach has been the starting point for the Digital European Cordless Telecommunications Standard (DECT) which is currently being developed using elements of a slightly earlier proprietary standard sometimes referred to as CT-3 and developed by the Ericsson Organisation as a wireless PBX product. Fig. 10.8 shows the basic frame structures as currently specified for DECT and CT-3.

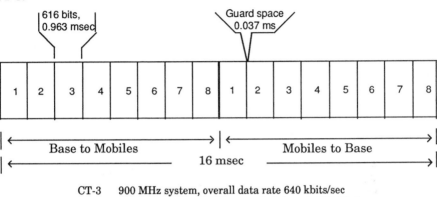

CT-3   900 MHz system, overall data rate 640 kbits/sec
channel spacing 1 MHz, speech data rate 32 kbits/sec

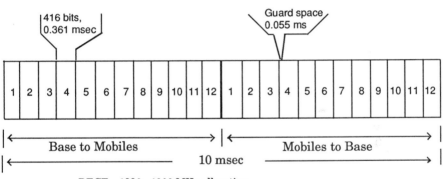

DECT   1880 - 1900 MHz allocation
overall data rate 1.152 Mbits/sec, channel spacing 1.728 MHz
speech data rate 32 kbits/sec

Fig. 10.8 TDMA frame composition in DECT and CT-3

The attractions of the TDMA approach in wireless office environments rests on the dual advantages of reduced numbers of base station installations for a given size of terminal population and on the possibility of allocating more than one time slot per frame to a single user wishing to operate at a higher bit rate than that required for a single speech circuit. This latter feature is significant in the context of interface compatibility with ISDN basic rates of B (64 kbits/s) and D (16 kbits/s) for future non-voice services.

    However, in the context of Telepoint services, these advantages are less significant, since in many situations one or two channels at a telepoint will satisfy demand and the added complexity of the multichannel TDMA is hard to justify. Further, at unattended sites, reliability of service tends to favour duplicated or triplicated single-channel-per-carrier base stations rather than single multi-channel ones.

Multicoupling of a number of single channel per carrier base stations onto a single transmit/receive antenna, might be perceived as a potential problem not shared by the multichannel TDMA approach, but the Telepoint requirement seeks only low power emissions from base stations and portables so that passive multicouplers using hybrid combiner techniques rather than cavity resonator filters, can be deployed without inhibiting frequency selection freedom in the self trunking type of operation described above for CT-2 CAI.

The relative merits of DECT/CT-3 and CT-2 plainly offer ample opportunity for debate on the choice of standard for new applications based on Telepoint and wireless PBX requirements. Further aspects are considered in the last chapters (14 and 15) which deal with evolutionary aspects of cellular and cordless systems as they begin to converge in personal communication networks.

Returning to the Telepoint service as distinct from air interface issues, the restrictions imposed by the need for hand portables to initiate calls and by the absence of handover capability in CT-2 CAI are clearly targets for further development in the emergence of a "Telepoint Plus" service. The fixed network infrastructure is acquiring increased intelligence with diffusion of ISDN technology and the means already largely exist for providing local registration possibilities for hand portables so that they can receive incoming calls from the network during a fixed location operation of any significant duration.

Plainly the basic telepoint service concept offers considerable scope for evolution into new services and corresponding opportunities for manufacturers to innovate in terminal and base station design. The prospect in view is now of a continuum of services being available to public subscribers ranging from the current CT-2 Telepoint level of service through Telepoint Plus, DECT and GSM based PCNs to full pan-European cellular capability. The potential market for the services in total is vast, the Telepoint/cordless segment is bound to be a significant proportion of this total.

10.9   References

1.   DTI Radiocommunications Agency (RA): 'Common air interface specification to be used for the interworking between cordless telephone apparatus including public access service', MPT 1375, May 1989

2.   DTI RA: 'Performance specification. Radio equipment for use at fixed and portable stations in the cordless telephone service operating in the band of 864.1 to 868.1 MHz', MPT 1334, Dec 1987

3.   BS 6833: 'Apparatus using cordless attachments (excluding cellular radio apparatus) for connection to analogue interfaces of public switched telephone networks', British Standards Institution, 1987

4.   DTI RA: 'Code of practice. Transmission of digital information over land mobile radio systems', MPT 1317, April 1981

Chapter 11

Type approval for equipment used in mobile radio services

Norman W J Lightfoot

11.1 Introduction

Equipment used in the Mobile Radio Services operate in a limited part of the frequency spectrum, which is determined by international agreement through the International Telecommunications Union. For a detailed description of the international regulatory arrangements see Skiffins [1]. Administration of this part of the spectrum at national level is controlled by Government Administration, which in the United Kingdom is the Department of Trade and Industry (DTI), Radio Department or Radiocommunications Agency. To ensure efficient use of the mobile radio spectrum, the Department of Trade and Industry control frequency assignments, conditions of use, equipment performance and ensure that mutual interference between users is acceptable and kept to a minimum.

The equipment used in the service operates on discrete radio frequency channels with narrow channel spacings and amplitude modulation and angle modulation systems are both in use. In the past, channel spacings as wide as 200 kHz were used, but have been progressively reduced with the increasing demand for private and public mobile radio. The channel spacings used today in the United Kingdom are 12.5 kHz in both the VHF and UHF frequency bands. There are some services such as paging and cellular radio systems which use wider channel spacings (25 kHz), and also several non-land mobile systems use wider spacings.

The performance limits set by the DTI Radio Department for mobile radio equipment are enforced by type approval testing procedures, which in conjunction with the need of the user, imposes high technological standards upon the equipment. To keep abreast of the technology, practical assessment of equipment performance is vital and the analysis by measurement provides the necessary link between theory and practice. The specification requirements for mobile radio equipment also imposes special requirements on instrument manufacturers, and to clarify the needs of the industry, it is useful to examine the important parameters of the receivers and transmitters used in the service, which are:

For receivers:
- Sensitivity
- Adjacent channel selectivity
- Co-channel rejection
- Blocking or desensitization
- Spurious response attenuation
- Intermodulation response

For transmitters:
- Radio frequency output power
- Frequency accuracy and tolerance
- Inter-transmitter intermodulation
- Adjacent channel power
- Conducted spurious emissions

11.2 Radiated measurement

In addition, equipment radiates both intentionally and unintentionally, defined by the characteristics.

- Receiver spurious emissions
- Transmitter spurious emissions
- Receiver radiated sensitivity
- Transmitter radiated RF power output

These and the previous characteristics do not give an absolute assessment of mobile radio equipment, but have been chosen as the most significant, particularly in terms of the type approval requirements which are used to check;

(1)   Parameters which effect the receivers performance in the presence of interfering signals from other mobile radio systems and communication services,

and

(2)   Parameters of the transmitter which can cause interference to other mobile radio systems and communication services.

There are a multitude of standards used throughout the world to assess identical parameters, each giving a different result. This unfortunately has the effect of adding to the cost and complexity of mobile radio equipment and in a likewise manner to the test instruments required to perform the measurements

11.3 European telecommunication standards

In Europe harmonisation in the methods of measurements for mobile radio equipment has been introduced as a result of the recommendations (TR/24) issued by the Conference of European Posts and Telecommunications administrations (CEPT). Industry, however, was not formally allowed to take part in the discussions which led up the CEPT recommendations. The work of CEPT in this area has now been transferred to the newly formed European Telecommunications Standards Institute (ETSI), which allows industry to participate on a more equal footing with the PTT administrations. The new standards, when agreed, will be formally issued as an ETS. Internationally a similar exercise is being carried out by the International Electrotechnical Commission (IEC) in their standards contained in IEC Publication 489.

Clearly if a single standard was adopted this would be beneficial to all concerned in the mobile radio services.

The measurements described in this paper are selected from those which are in the most common use and also serve as a guide to the problem which can arise when carrying out practical tests. The measurements are confined to angle modulated equipment, but are equally applicable to amplitude modulated equipment if the test modulation is altered to suit this emission.

11.4 Receiver measurements

11.4.1 Measurement bandwidth

To obtain consistent results and carry out meaningful comparisons of receiver performance the audio frequency bandwidth used in the measurements must be controlled. Propriety instruments used to monitor the audio output of a receiver, often have audio bandwidths far in excess of that of the receiver which leads to measurement errors.

The CEPT administrations have overcome this problem by recommending the use of the CCITT telephone weighting network P53. This network is acceptable for most receiver measurements except for characteristics such as audio response and audio distortion.

The IEC have recommended an audio measurement bandwidth not exceeding 10 kHz. This is more acceptable and can be used in all measurements of the audio characteristics for the receivers used in the mobile services, where the highest audio frequency does not exceed 3000 Hz.

11.4.2 Sensitivity

The receiver sensitivity is measured in terms of the level of a radio frequency signal applied to the input of a receiver at its nominal input frequency which produces a specified signal-to-noise ratio at the output of the receiver. This is a straight forward measurement and is of great importance as it is used as a reference for the majority of other measurements carried out on the receiver, and is also used in the planning of mobile radio systems.

One method commonly used to define the signal-to-noise ratio of the receiver is the SINAD ratio. This is expressed as

$$\text{Signal-to-Noise Ratio (SINAD)} = \frac{\text{Signal} + \text{Noise} + \text{Distortion}}{\text{Noise} + \text{Distortion}}$$

The value of the SINAD ratio normally used by CEPT is 20dB which must be achieved for a specified RF signal at the input to the receiver.

The test arrangement of the measurement of SINAD is shown in Fig. 11.1.

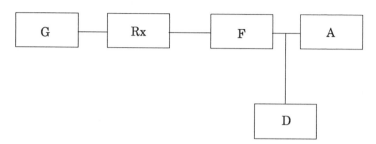

G   =   Radio frequency (RF) signal generator
Rx =   Receiver under test
F   =   Audio filter
A   =   Audio frequency output power meter
D   =   Distortion factor meter

Fig. 11.1 Test arrangement for the measurement of
receiver sensitivity

The measurement is performed in the following manner; apply an RF signal, modulated with a test tone 1 kHz at 60% of the maximum "system deviation") at a level of 1 mV to the input of the receiver at its nominal frequency. The audio output of the receiver is adjusted to a convenient reference, normally 50% of its rated output power.

The test tone is then removed by use of a notch filter; in normal practice a distortion factor meter is used. The output of the RF signal generator is then reduced until the ratio of the reference audio output power to the residual output power is 20 dB. (This is equivalent to an overall audio distortion of 25%).

Although sensitivity is a simple test, the measurement is prone to significant inaccuracies. Rudkin [2] has shown that errors of greater than 6 dB can be encountered and are mainly caused by the following measurement parameters.

(1)   Impedance matching

The nominal input impedance of a receiver used in the mobile service is 50 ohms, but in practice the impedance is not 50 ohms.

This difference in the impedance coupled with any inaccuracy in the output impedance of the RF signal generator introduces error into the measurement. This matching problem has largely been overcome by stating the input voltage to the receiver in terms of the open circuit voltage (emf) of the signal generator or the equivalent voltage which would be developed across an impedance equal to the nominal impedance of the receiver, which results in a voltage equal to half the emf.

Errors due to the voltage standing wave ratio (VSWR) of the signal generator can be significant. A voltage standing wave ratio of less than 1.2:1 is necessary to reduce error to an insignificant level.

(2)   Signal generator performance

The requirements of the signal generator are stringent, particularly in the following areas.

Radio frequency tolerance and setting accuracy. As mentioned previously, channel separations of 12.5 kHz are used and receivers used in this service have IF bandwidths of 8 kHz. Therefore, it is necessary to set the frequency within ± 100 Hz of the receiver operating frequency with a short term frequency stability better than $\pm 1 \times 10^{-7}$.

Attenuation and radio frequency leakage are common sources of error.

The accuracy of the signal generator output level is largely dependent on the attenuator which controls its output.

In the high quality signal generators now available this error has been reduced to ± 1 dB at any output level setting.

Mobile radio receivers have sensitivities of less than 0.5 µV (emf) and therefore leakage of a signal generator is critical. The effect of this leakage is that optimistic sensitivity figures are obtained. A simple test to prove this point is to move the receiver further away from the signal generator, if this necessitates an increase in the generator output, then the generator suffers from leakage. To prove the point conclusively, the receiver should be placed in a Faraday Cage to isolate the generator from the receiver under test.

Modulation setting errors are another contributing factor, but as generators with an accuracy of ± 5% are now available, the errors in this area have been reduced to an acceptable level.

Selecting the best performance in all the parameters mentioned, has made it possible to reduce the overall measurement error to less than 2 dB.

11.4.3   Receiver selectivity

The characteristic of the receiver is defined as, the ability of the receiver to discriminate between the wanted input signal and the unwanted signal, the wanted signal and the unwanted signals being present simultaneously at its input.

The parameters of the receiver included under this heading are adjacent channel selectivity spurious response, attenuation, blocking or desensitization, co-channel rejection and intermodulation response.

The test arrangement required to carry out these measurements is shown in Fig. 11.2

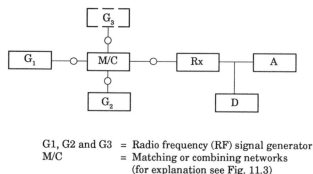

G1, G2 and G3 = Radio frequency (RF) signal generator
M/C   = Matching or combining networks
                       (for explanation see Fig. 11.3)
Rx   = Receiver under test
A   = Audio frequency output power meter
D   = Distortion factor meter

Fig. 11.2 Test arrangement for the measurement of the
receiver selectivity characteristics

11.4.3.1   *Adjacent channel selectivity*

The adjacent channel selectivity is a measurement of the effect on the signal-to-noise at the receivers output due to a wanted signal in the presence of an unwanted signal in the adjacent channel.

The signal is performed by connecting two radio frequency generators to the RF input of the receiver via a combining network. The audio output of the receiver is connected to a distortion factor meter and an audio monitoring device, which includes a CCITT telephone weighting network P53. Referring to Fig. 11.2; let G_1 represent the wanted signal and G_2 represent the unwanted signal. The combining network in this instance is a three port device, each port having an impedance of 50 ohms and a resulting network attenuation of 6 dB. Examples of combining networks are shown in Fig. 11.3.

In the absence of the unwanted signal G_2, the frequency of the signal generator G_1 is adjusted to the nominal frequency (fc) of the receiver with its output modulated in the same manner as for the sensitivity test. The output level of G_1 is then adjusted to a specified level at the input to the receiver.

The specified level of RF signal at the input of the receiver is chosen by the regulatory authority and is the level of signal used for frequency planning purposes where a key consideration in the allocation and re-use of frequencies is the protection of a specified field strength, which relates to a specified RF signal level at the input to the receiver. The limit chosen by CEPT is 6 dBμV.

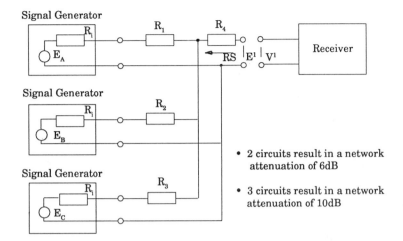

Fig. 11.3 Radio frequency signal generator combining networks

A signal modulated with 400 Hz at 60% of the maximum system deviation is then applied to the receiver from generator G_2 simultaneously with G_1. The frequency of generator G_2 is adjusted to the upper adjacent channel (fc + ∆f), where ∆f is the channel spacing, (e.g. 12.5 kHz) and its output level is increased until the SINAD ratio of 14dB is obtained.

The measurement is repeated with generator G_2 adjusted to the lower adjacent channel (fc - ∆f). The ratio of the output level of signal generator G_1 to the level of signal generator G_2 is the adjacent channel selectivity of the receiver. The lower of the two values measured is the adjacent channel selectivity.

This measurement is normally repeated at temperature and voltage extremes.

Examining the performance requirements of the test instruments, the two signal generators used can have a significant effect on the results. Signal generator G_1, which represents the wanted signal must have a performance equal to that of the signal generator described for the sensitivity measurements. Signal generator G_2 representing the adjacent channel signal will impose the greatest limitation of the measured results, due to the level of adjacent channel noise present in its own output signal. Referring to Fig. 11.4, the attenuation of the noise power of the unwanted signal which extends over the IF passband of the receiver must be greater than the receiver adjacent channel selectivity performance.

The majority of signal generator manufacturers state this attenuation in terms of the noise power measured in a 1 Hz bandwidth (dB/Hz) relative to the output power of the signal generator. This value of the attenuation must be modified to take into account the total noise power which will appear in the IF passband of the receiver.

Assuming that the spectrum of noise produced by the signal generator in the region of the receiver IF passband is flat, a simple calculation can be used to estimate the limit of the selectivity measurement due to this signal generator noise spectrum. Before proceeding with the calculation, another characteristic of the receiver must be included, namely the Co-Channel Rejection Ratio which will be in the order of 10 dB for a receiver of this type.

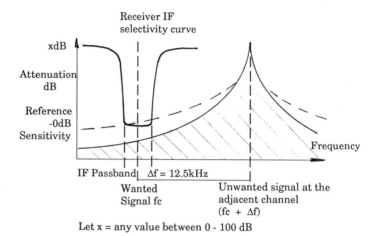

Let x = any value between 0 - 100 dB

Fig. 11.4   Effect of noise characteristic of adjacent channel signal generator

If we take the example of a 12.5 kHz channel spaced receiver, the IF passband to be considered is 8 kHz, and assuming that the adjacent channel performance of the receiver is 70 dB,

Therefore,   The Minimum attenuation of the noise power (dB/Hz) with respect to the output power of the signal generator at its nominal frequency

= (Adjacent channel selectivity ratio) +

10 log$_{10}$ 8000

+ (Co-channel rejection ratio)

= (70 + 39 + 10) dB/Hz

= 119 dB/Hz

Because the receiver IF passband is centred on the adjacent channel frequency, allowance must be made for the portion (half) of the IF passband which extends towards the wanted signal generator carrier frequency (f). Using the above example, the attenuation of 119 dB/Hz is necessary at a frequency fc ± (12.5 - 4) kHz = fc ± (8.5) kHz.

The adjacent channel selectivity of present day receivers (12.5 kHz channel separation) is approaching 80 dB. This is difficult to measure, because of the performance limitation with regard to the adjacent channel noise in currently available signal generators.

11.4.3.2 *Co-channel rejection ratio*

The test arrangement and method of measurement for this receiver characteristic is similar to the adjacent channel selectivity test, except that the unwanted signal frequency at the output of G_2 (Fig. 11.2) is adjusted to within plus or minus 3000 Hz of the nominal frequency of the receiver. The level of this signal is then adjusted until a SINAD ratio of 14 dB is obtained.

The co-channel rejection ratio is the ratio of the output level of signal generator G_1 to signal generator G_2.

The requirements of both signal generators in this test are identical to that necessary for the receiver sensitivity measurement.

Co-channel rejection ratios of 8 to 10 dB are typical for current day receivers.

11.4.3.3 *Blocking or desensitization*

The blocking or desensitization of the receiver can be considered as an extension of the adjacent channel measurement.

The main differences are the frequency offsets of the unwanted signal, which is normally equal or greater than 1 MHz, and the measured effect on the receiver audio output. The level of the unwanted signal is adjusted until either the following occur.

(a)   A reduction in the audio output power of the receiver by 3 dB

  or

(b)   A SINAD ratio of 14 dB is obtained, which ever occurs first is the point at which receiver blocking level is measured.

Blocking levels of 90 dB/μV or higher are normal in receiver specifications.

This measurement imposes the same stringent noise power attenuation performance for the signal generator as with adjacent channel measurement, but in this case, the unwanted signal frequency offset is greater. The noise power attenuation of the signal generator must be greater than 133 dB/Hz to measure a blocking level of 90 dB/μV.

11.4.3.4 *Spurious response attenuation*

Receivers designed for use in the mobile services normally employ single or double superheterodyne techniques. This necessitates the use of a mixer to generate the intermediate frequency (IF). Two frequencies are applied to the mixer input, the sum or difference of these frequencies is equal to the required intermediate frequency.

One of these frequencies is generated within the receiver (mixer injection frequency) and the other is the radio frequency signal applied to the input of the receiver.

Spurious responses of the receiver are caused by unwanted signals appearing at the input of the receiver, mixing with harmonics of the internally generated frequency(s) such that their sum or difference is equal to the intermediate frequency of the receiver.

The frequency of spurious response for this type of receiver can be calculated in the following manner,

$$f_1 = nf_2 + f_{IF}$$

where f_1 = the unwanted (spurious) frequency
          f_2 = Local oscillator frequency
          f_{IF} = Intermediate frequency
          n = an integer

The above equation applies to a single superheterodyne receiver. If a second intermediate frequency is used then the equation would need to be modified.

A two signal method of measurement is used to assess the spurious response attenuation of the receiver. The measurement arrangement is as shown in Fig. 11.2. The measurement is again carried in a similar manner to the adjacent channel selectivity, but here the unwanted signal frequency is set to that corresponding to a spurious response frequency.

The wanted signal is adjusted to a specified level at the input to the receiver. The level of the unwanted signal is then adjusted until a SINAD ratio of 14 dB is obtained.

The ratio of the levels of the wanted signal to the unwanted signal is the measure of the receivers spurious response attenuation.

The measurement is repeated for each spurious response of the receiver.

The CEPT recommendation requires that all spurious responses from the lowest intermediate frequency of the receiver to an upper frequency of 1 GHz are measured. An attenuation of 70 dB with respect to the specified level at the input to the receiver is the limit value.

In this measurement the performance of the signal generator (G_1) used as the wanted signal is identical to that necessary for the receiver sensitivity measurement. The signal generator which represents the unwanted signal, must have an output spectrum which is substantially free from .spurious signals over the measurement frequency range. The broadband noise floor of the output signal must have an attenuation with respect to its nominal frequency that is greater than the spurious response attenuation of the specification limit.

11.4.3.5 *Intermodulation response attenuation*
This characteristic can be defined as ability of the receiver to discriminate between wanted input signal and more than one unwanted input signal with specific frequency relationships to the wanted signal frequency.

The test arrangements required to carry out the measurement is shown in Fig. 11.2. Three radio frequency generators are connected to the receiver radio frequency input via combining network. Referring to Fig. 11.2, let G_1 represent the wanted signal and G_2 and G_3 the unwanted signals. The combining network required for this measurement is a four port device as described in Fig. 11.3. Each port has an impedance of 50 ohms and the network has an attenuation of 10 dB. In the absence of the two unwanted input signals, the frequency of signal generator G_1 is adjusted to the nominal frequency (fc) of the receiver, with its output modulated in the same manner as for the sensitivity test. The output level of generator G_1 is then adjusted to a specified level at the input to the receiver. An unmodulated signal is then applied to the input of the receiver from signal generator G_2, the frequency of this generator is set to the upper adjacent channel (fc + Δf), where Δf is the channel separation, e.g. 12.5 kHz. Simultaneously a signal modulated with 400 Hz at 60% of the maximum system deviation is applied to the input of the receiver from generator G_3. The frequency of generator G_3 is adjusted to the upper alternate channel (fc + 2Δf).

The output levels of signal generator G_2 and G_3 shall be equal and at a level which reduces the SINAD ratio to 14 dB. Slightly adjust the frequency of one of the unwanted signals G_2 or G_3 to obtain the maximum reduction in the SINAD ratio. Finally adjust the levels of G_2 and G_3 to re-establish the SINAD ratio of 14 dB.

The measurement is repeated with generator G_2 adjusted to the lower adjacent channel (fc - Δf) and generator G_3 adjust to the lower alternate channel (fc - 2Δf). The intermodulation response attenuation is the ratio of the level of the wanted signal to one of the equal levels of the two unwanted signals. Receivers used in the mobile service are required to have intermodulation response attenuation ratios in excess of 70 dB.

The performance of the signal generator (G_1) used as the wanted signal is identical to that necessary for the receiver sensitivity measurement.

The two signal generators $(G_2$ and $G_3)$, used as the unwanted signals. This is due to the high level of signal at their output, and insufficient isolation in the combining network. To verify this point, insert a calibrated stepped attenuator between the combining network and the receiver under test. If no intermodulation is present between the signal generators, then for each dB of attenuation added between the combining network and receiver, all three signal generators $(G_1, G_2,$ and $G_3)$ output levels will need to be increased by the same amount. If intermodulation is taking place between the signal generators, then the two unwanted signal output levels will need increasing by a smaller amount than the attenuator step. This situation can be improved but replacing the resistive combining network, with a hybrid coupler, this will improve the isolation between the signal generators, although the tests must be repeated to establish the measurement limit.

If	$f_2^{'}$	=	fc $\pm$ Δf
and	f_3	=	fc $\pm$ 2Δf
then	fc	=	$2f_2 - f_3$
where	fc	=	the frequency of the wanted signal (G_1)
	f_2	=	the frequency of the nearer unwanted signal (G_2)
	f_3	=	the frequency of the more remote unwanted signal (G_3)
and	Δf	=	channel separation, e.g. 12.5 kHz

It is now becoming the accepted practice in some specifications to measure higher orders on intermodulation.

11.5 Transmitter measurements

11.5.1   Carrier power

The carrier power is the average power supplied to a transmission line by a transmitter during one radio frequency cycle in the absence of modulation. This definition is in conformity with ITU Regulations.

In a practical measurement the transmission line is replaced by a test load or Radio Frequency (RF) Power Meter.

The measurement is made connecting an RF power meter to the radio frequency output terminals of the transmitter. Operated the transmitter and note the RF power reading in watts.

RF power meters are readily available with an accuracy of 5% at full scale reading. The majority of specifications require an overall accuracy equal to, or less than, 10%.

Apart from the accuracy of the power meter, there are two other fundamental requirements which are:

1)   The impedance of the RF power meter must be equal to the nominal impedance of the transmitter output circuit. For mobile radio equipment this is 50 ohms.

2)   A good quality co-axial cable with a characteristic impedance of 50 ohms must be used. Good RF connectors must be used with the cable to ensure that the vswr is as low as possible.

11.5.2   Frequency accuracy and tolerance

The frequency accuracy of transmitters used in the mobile services is extremely important, bearing in mind the constraints of operating on discrete channel frequencies with narrow channel spacings. If we take an example of a transmitter operating in the top end of the VHF frequency band with a channel spacing of 12.5 kHz, the transmitter frequency will need to be adjusted to within 100 Hz of the assigned frequency. This same constraint applied to equipment operating in the UHF frequency band. Transmitters in this service must also have good frequency tolerance, at voltage and temperature extremes. In the United Kingdom, the temperature limits are -10°C to +55°C and the voltage limits vary depending on the type of power supply used, e.g. AC power supply in the case of a fixedstation equipment and DC power supply for the many types of batteries for portable equipment.

The equipment tolerance limits for VHF equipment is generally in the order of $\pm$ 10^{-5} and for UHF equipment $\pm$ 5 x 10^{-6}. These vary slightly in different countries throughout the world.

To measure the frequency of a transmitter, a frequency counter is connected to the radio frequency output via a suitable attenuator. The attenuator must be capable of dissipating the transmitter power output and the amount of the attenuation must be such that the input circuits of the frequency counter are not overloaded. The frequency tolerance of the counter must be at least an order better than that of the transmitter.

The most important factor in relation to the transmitter frequency tolerance is the potential interference problem that could arise, as any movement in frequency will bring the transmitter carrier and its sideband products closer to the next operating channel.

11.5.3 Inter-transmitter intermodulation

Inter-transmitter intermodulation is becoming an ever increasing problem due to the growth of communal (shared) sites. These sites house several fixed stations (transmitters and receivers) whose antennas are often in close proximity to each other, i.e. see Chapter 12.

The result of this congestion is that signals which are radiated from one antenna are coupled into another antenna whose transmitter is operating. This means that signals other than its own will appear in the output stages of a transmitter, and any non-linearities in the transmitter power amplifiers will cause intermodulation products to be produced and re-radiated. Circulators are used to reduce this problem, but the basic transmitter must have a good performance with respect to this characteristic.

The inter-transmitter intermodulation product is measured as the ratio of the intermodulation product level to the level of the unwanted signal. This ratio is sometimes known as the "Transmitter Intermodulation Conversion Loss".

The measurement of this characteristic is carried out with the test arrangement set up as shown in Fig. 11.5. The radio frequency (RF) signal generator (G) represents the unwanted signal from another transmitter. The output level of this generator is adjusted to a value which produces a specified "Coupling Factor" between the two transmitters, e.g. 30 dB, 40 dB, etc., with a known frequency relationship. The coupling factor is the ratio of the output power of the wanted transmitter to the power appearing at its RF output terminals from the unwanted transmitter. Initially, the RF power output transmitter with the test load (2) and adjust the frequency of the RF signal generator (G) to the frequency of the adjacent channel (e.g. 12.5 kHz) above the nominal frequency of the transmitter. Adjust the output of this RF generator to obtain the required coupling factor.

Next replace the test load (2) by the transmitter. Operate the transmitter and adjust the spectrum analyser (3) frequency and IF bandwidth, to display the transmitter output, the unwanted signal and any resulting intermodulation products.

Transmitter intermodulation conversion loss ratios of between 5 dB and 10 dB can be expected, depending on the type of transmitter. The CEPT limit is stated as the ratio of the power level for the transmitter to the power level of the intermodulation component and requires a minimum of 42 dB for any intermodulation component.

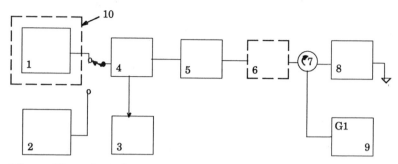

1 Transmitter under test
2 Test load including radio frequency
    power meter
3 Selective measuring device
    (Spectrum Analyser)
4 Coupler
5 Line stretcher

6 Attenuator if needed
7 Circulator
8 Test load
9 Radio frequency signal generator
10 Faraday cage

Fig. 11.5   Test arrangement for the measurement
of inter-transmitter intermodulation

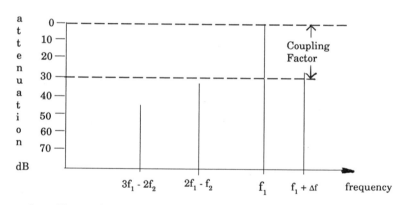

f_1 =   The wanted transmitter carrier
$f_1 + \Delta f$ =   The unwanted transmitter frequency
Δf =   Channel separation e.g. 12.5kHz

$2f_1 - f_2$ = Third order intermodulation product
$3f_1 - 2f_2$ = Fifth order intermodulation product

Fig. 11.6   Transmitter intermodulation products

Fig. 11.6 illustrates the intermodulation product components. The line stretcher is now adjusted to obtain the highest level of intermodulation products. Repeat the procedure used to establish the coupling factor before recording any results.

Repeat the measurement with the unwanted signal set to the adjacent channel below the nominal frequency of the transmitter.

Special attention must be given to the following factors, to reduce measurement errors to a mimumum

a)   Sufficient isolation between the transmitter and the unwanted signal isimportant to keep any coupling due to radiated signals to a level which will not affect the results. The use of a Faraday cage will overcome this problem, but care is still needed with the physical positioning of interconnecting cables, the signal generator, attenuator, line stretcher etc.

b)   Measurement linearity and dynamic range must be checked before proceeding with the measurement. First display the RF carrier of the transmitter in the centre of the spectrum analyser screen, with its peak on the 0 dB line. The transmitter carrier power must be attenuated such that is within the linear working range of the RF input circuits of the spectrum analyser. The unwanted signal frequency is then adjusted to the adjacent channel and its level varied in 10 dB steps from -10 dB to -70 dB with respect to carrier power of the transmitter. At each step, check that the displayed signal changes by the same amount.

11.5.4   Adjacent channel power

The adjacent channel power of a transmitter is that part of the total output of the transmitter under defined conditions of modulation, which falls within a speci
Special attention must be given to the following factors, to reduce measurement errors to a minimum:

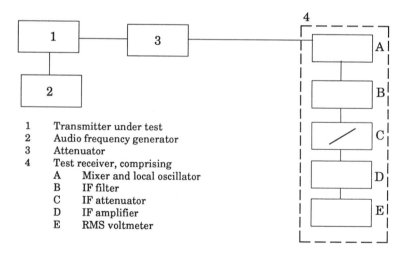

1	Transmitter under test
2	Audio frequency generator
3	Attenuator
4	Test receiver, comprising
	A Mixer and local oscillator
	B IF filter
	C IF attenuator
	D IF amplifier
	E RMS voltmeter

Fig. 11.7   Test arrangement for the measurement
of transmitter adjacent channel power

This characteristic of the transmitter has a significant effect on frequency planning with respect to channel allocation and the geographical separation of repeated channel frequencies. The test arrangement illustrated in Fig. 11.7 is used for the measurement. Operate the transmitter at its normal rated power output, with defined conditions of modulation (e.g. CEPT and IEC use 125 Hz at a level which is 20 dB greater than the level which is produced 60% of the maximum system deviation). Adjust the attenuation (3) such that the signal level from the transmitter is within the linear portion of the test receiver's operating range. Next adjust the IF attenuation (4C) to obtain a convenient reference level on the rms voltmeter (4) but which is 10 dB greater than the receiver noise output. Adjust the frequency of the test receiver to the upper adjacent channel and re-adjust the attenuator (3) until the reference level as set previously on the rms voltmeter (4D) is obtained.

The adjacent channel power can be calculated from the difference in the attenuator (3) values (dB) which were set at the transmitters nominal frequency and at the adjacent channel frequency.

The measurement is repeated with the test receiver at the lower adjacent channel frequency.

The test receiver's performance will limit the value of adjacent channel power which can be measured. The important characteristics of this receiver are:

(1)   The adjacent channel selectivity;
       this limits the ratio of the transmitter carrier power to the power in the adjacent channel, which can be measured.

(2)   The test receiver must be screened, such that its performance in the presence of a high level radiated signal (the transmitter under test) is not affected. If this is found to be a problem, then the transmitter must be isolated from the test receiver, by using a Faraday cage.

(3)   The characteristics of the IF filter (4B) must correspond to that normally used for the channel spacing in which the transmitter operates. CEPT and the IEC have defined this characteristic.

11.5.5   Conducted spurious emissions

These are radio frequency components which are present at the output of the transmitter at frequencies other than the wanted carrier frequency and its associated wanted modulation components. These spurious components include, harmonics of the oscillator frequency, used to generate the carrier frequency, harmonics of the carrier frequency and components due to parasitic oscillations. Specification limits are stringent for this transmitter characteristic. A power of $0.25\mu W$ is the value set in the CEPT Recommendations. The test arrangement for this measurement is shown in Fig. 11.8.

Operate the transmitter at its normal carrier power. Adjust the frequency of the selective voltmeter to the lowest frequency spurious component of the transmitter and note the frequency and the deflection of the selective voltmeter.

Replace the transmitter by the RF signal generator (2) and adjust its frequency to that of the selective voltmeter. Finally, adjust the output level of the signal generator to obtain the same deflection of the selective voltmeter as the transmitter. The amplitude of the spurious emission is equal to the output signal level of the signal generator (2). Repeat the measurement for each spurious emission over a specified frequency range. When carrying out the measurement for the harmonics of the wanted carrier frequency a filter will need to be inserted between the test load (3) and the selective voltmeter (5) to remove the carrier component in order to overcome any non-linearities in the measurement due to the presence of a high level signal at the input of the selective voltmeter.

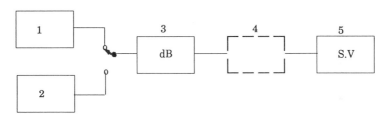

1	Transmitter under test
2	Radio frequency (RF) signal generator
3	Test load attenuator
4	Band - rejection or high pass filter if required
5	Selective voltmeter (Spectrum analyser or radio frequency interference receiver)

Fig. 11.8 Test arrangement for the measurement of
conducted spurious emissions

There are proprietary radio frequency interference (RFI) meters available, with level measurement accuracies of ± 1 dB. If this type of instrument is used it is not necessary to use the substitution method described above.

A careful approach to this measurement is necessary as comparatively small signals are being measured in the presence of high level signals (the transmitter carrier power). The RFI meters mentioned previously normally have good radio interference protection, but if there are any doubts in the measurement results, use a Faraday cage to isolate the transmitter from the selective voltmeter. High quality attenuators must be used with a voltage standing wave ratio of 1.2:1 or less over the frequency range of the measurement. The coaxial cables and connectors used must also be high quality.

11.6 Radiated measurements

There are two basic applications where radiated measurements need to be used:

(a)   To measure spurious emissions emitted from the chassis or case of the equipment. The limits for this measurement are stated as a field strength in μV/metre or as an absolute power in watts.

The measurement of radiated spurious emissions are applicable to all types of equipment used in the mobile services, fixed stations, mobiles, portables and pagers etc.

and

(b)   To measure transmitter and receiver characteristics for equipment which has an integral antenna, where no direct connection can be made to the equipment without disturbing its normal operation and performance. Examples of the type of equipment are pagers and portables.

This type of measurement is normally carried out on an open radio frequency test site. To obtain consistent results it is necessary to define the key parameters of the test site, which are:

1   The test site shall have a specified measuring distance between the equipment under test and the measuring antenna. Measuring distances used for test sites vary throughout the world, the most commonly used are the 3 metre, 10 metre and 30 metre sites. In practice, results obtained from these different sites can vary as much as 10 dB.

2   The test site shall be on level ground having uniform electrical characteristics and be free from reflecting objects over as wide an area as possible to ensure that extraneous electromagnetic fields do not affect the accuracy of the test results.

3   The position of the equipment under test must be specified. Normally the equipment is placed on a pedestal made of non-conducting material positioned 1.5 metres above the ground.

4   The measuring antenna must be suitable for the reception of linearly polarised waves. It may consist of a half-wave dipole, the length of which is adjusted for the frequency concerned, often calibrated log-periodic antennas are used to cover a wide frequency measurement range.

5   An auxiliary antenna which is used to replace the equipment under test during part of the measurement. This antenna is similar to the measuring antenna.

6   A well shielded radio frequency generator and its associated output cable.

7   A selective measurement device which may be either a selective voltmeter, a spectrum analyser or a calibrated field strength meter.

The example chosen to describe the radiated measurement is a 30 meter test site. The measurement arrangement for this test site is shown in Fig. 11.9.

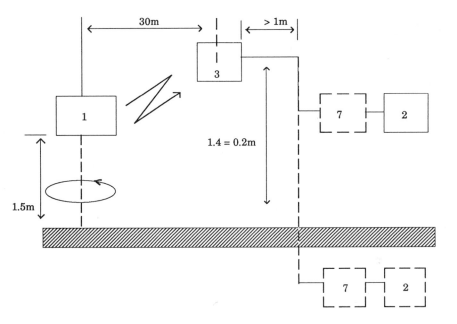

Fig. 11.9 30m test site arrangement for measurements

1 = Equipment under test
2 = Selective measuring device
3 = Measuring antenna
4 = Auxiliary antenna
5 = Radio-frequency signal generator
6 = Selective measuring device

7 = Optional calibrated attenuator
for measuring the effective
radiated power, or fundamental
oscillation rejection filter for
measuring non-essential
emissions

11.6.1 Radiated spurious emissions

These are radio frequency components radiated from the chassis or case of the
equipment. These spurious components are applicable to both the transmitter
and receiver. They include; 1) harmonics of the oscillator frequency used to
generate the transmitter carrier frequency, harmonics of the carrier frequency
and components due to parasitic oscillators. The wanted carrier frequency and
its associated wanted modulation components are excluded, and 2) harmonics of
the receiver local oscillator(s) frequency.

The method of measurement for radiated spurious emissions is performed in
the same manner for both the transmitter and the receiver.

Referring to Fig. 11.9:

a)   Place the equipment under test on the test platform or pedestal which is
        1.5 metres above the ground. If the equipment has an antenna terminal,

it shall be terminated in a test load having an impedance equal to the nominal radio-frequency input impedance.

b)   Operate the receiver.

c)   Identify the frequencies of the significant spectral radiated spurious components by using a selective measuring device. If necessary, closely couple it to the receiver under test.

d)   Adjust the measuring antenna (if adjustable) to the correct length for the frequency of one of the significant spectral radiated spurious components identified in step d).

e)   Tune the selective measuring device to the considered significant spectral radiated spurious component.

Note: deactivate the receiver under test to verify that it is the origin of the radiated spurious component in step f).

f)   Position the measuring antenna for vertical polarization.

g)   Rotate the equipment under test to obtain the maximum indication on the selective measuring device.

h)   When required by the test site description, raise and lower the measuring antenna to obtain the maximum indication on the selective measuring device.

i)   Repeat steps h) and i) until no further increase occurs. Note the frequency and the maximum indication.

j)   Position the measuring antenna for horizontal polarization and repeat steps h), i) and j).

k)   Repeat steps e) to k) until the levels have been measured for all significant spectral radiated spurious components identified in step d).

l)   Replace the receiver under test by a vertical auxiliary antenna, refer to Fig. 11.10.

m)   Select one of the significant spectral radiated spurious components measured in step j) and adjust the frequency of the radio-frequency signal generator to its frequency.

n)   Adjust the length of the auxiliary antenna (if adjustable) for the considered frequency.

o) Adjust the length of the measuring antenna (if adjustable) for the considered frequency.

p) Tune the selective measuring device to the considered frequency.

q) Position the measuring antenna for vertical polarization.

r) Adjust the output level of the radio-frequency signal generator to provide an indication on the selective measuring device.

s) When required by the test site description, raise and lower the measuring antenna to provide a maximum indication on the selective measuring device.

t) Re-adjust the output of the radio-frequency signal generator to obtain the same value of indication as noted in step j). Note the output level of the radio-frequency signal generator and its frequency.

u) Repeat steps s) to u) with the auxiliary and measuring antennas horizontally polarized.

v) Repeat steps n) to v) for the remaining significant spectral radiated components measured in step j).

w) Calculate the power available to the auxiliary antenna from the values noted in step u) taking into account the different settings of the attenuator, the gain of the auxiliary antenna and the cable loss between the radio-frequency signal generator and the auxiliary antenna.

The power of the radiated spurious component is defined as the largest of the values thus calculated.

The measurement of the transmitter spurious emissions are identical to above, but with the transmitter operating.

The limit values quoted by CEPT for spurious emissions are, 0.25µW (100 kHz to 1000 MHz), 1µW (1000 MHz to 2000 MHz) for transmitters, and 2 nanowatts (100 kHz to 1000 MHz), 20 nanowatts (1000 MHz to 2000 MHz) per receiver.

11.6.2 Transmitter radiated RF power output

The measurement of the transmitter radiated RF power output is performed in the same manner and radiated spurious emissions except that the measurement is only carried out at the carrier frequency(s) of the transmitter. This measurement is normally only carried out on equipment with integral antenna.

11.6.3   Receiver radiated sensitivity

Use the test site shown again as in Fig. 11.10 for this measurement;

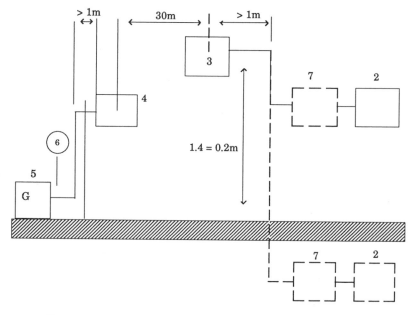

Fig. 11.10   30m test site arrangement for calibration
(For equipment see Fig. 11.9)

(a)   Place the equipment under test on the test platform or pedestal which is 1.5 metres above the ground and operate the receiver.

(b)   Apply the standard input signal from the radio-frequency signal generator to the measuring or transmitting antenna (3).

(c)   Increase its level to a high value and ensure that the receiver squelch is opened.

(d)   Orientate the equipment so that a face, specified by the manufacturer, is **normal to the direction of the measuring or transmitting antenna.** This orientation is the reference for the measurement of the azimuth.

(e)   In the case of an adjustable squelch, adjust it to open using an input-signal level which provides a signal-to-noise ratio between 10 dB and 20 dB, as estimated by a listening test. Retain this squelch setting.

(f)   Reduce the output level of the radio-frequency signal generator to close the squelch. Then increase the level until the squelch just opens, and record this level in μV.

(g)   Rotate the equipment under test successively in seven steps of 45° and record the output level of the radio-frequency signal generator required to just open the squelch at each of the azimuthal positions.

(i)   If the results indicate that at a particular azimuth angle the radio-frequency signal generator output level is significantly less than at other angles, determine the minimum radio-frequency signal generator output level required to open the squelch, by proceeding as follows:

        In the vicinity of the assumed azimuth for the minimum radio-frequency output level, select smaller azimuth rotation angles, for example 15°, and for each azimuthal position, determine the output level of the radio-frequency signal generator required to just open the squelch. Record thelowest level.

(j)   To calculate the actual field strength in μV/m around the equipment under test on the test site using the radio-frequency signal generator level recorded in step i), the test site must be calibrated in the following manner;

11.7   Calibrating the test site

1)   Connect the equipment as shown in Fig. 11.10. With the measuring or transmitting antenna oriented to provide the polarization intended of the receiver under test.

2)   Adjust the frequency of the radio-frequency signal generator to the operating frequency of the receiver.

3)   Tune the selective measuring device to the operating frequency of the radio-frequency signal generator.

4)   Adjust the output level of the radio-frequency signal generator to produce a reading of 100 μV/m (40 dB (μV/m)) on the selective measuring device. Record the output of the radio-frequency signal generator in μV.

5)   The field strength for other values of the radio-frequency generator output is given by:

$$\text{field strength} = \frac{\text{new output level}}{\text{output level recorded in step 4)}} \times 100 \ \mu/M$$

field strength = 40 + 20 log (new output level in μV) in dB (μV/M)
                  - 20 log (output level μV recorded in step 4)

Note   The calibration is only good for the frequency, antenna, polarization, and antenna position used in the calibration procedure. If any of these change, the site should be recalibrated.

The sensitivity of a receiver is normally measurement in terms of the RF input signal to achieve a specified signal-to-noise. In the example described earlier for receiver sensitivity a signal-to-noise of 20 dB SINAD was used. To carry out this measurement on a radiation test site is difficult, and therefore the measurement has been restricted to the level of signal which opens the squelch. To measure the signal-to-noise a radio frequency coupling device (RFCD) or test fixture can be used, although when this is used the following precautions must be taken to ensure that:

-       the receiver is adequately shielded from electromagnetic disturbances;

-       the attenuation of the coupling between the radion source and the receiver being measured is sufficiently low, stable and consistent throughout the measuring frequency range.

The couping loss depends on the particular measuring arrangement, the frequency being used, and the receiver being measured. Normally it is not precisely measured, as it will only be useful for a particular measuring arrangement and frequency.

The coupling loss, however, must be sufficiently low so that the output power requirements of the signal generators used in this standard will not exceed the power output capability of commercially available signal generators.

To ensure measurement repeatability, and RFCD which includes the following shall be used in the measuring arrangement:

-       a radiating element;

-       A radio-frequency input terminal connected to the radiating element through a transmission line;

-       a means to ensure that the input impedance of the RFCD is the same as the impedance of the transmission line from the radio-frequency signal generator;

-       a means for positioning the receiver being measured in a precise, repeatable and stable manner;

-       a means to ensure that the presence of the person making the measurement will not affect the results.

The RFCD shall also have the following characteristics.

-       *a* coupling loss between the radio-frequency input terminal and the receiver being measured of less than 30 dB;

-       a coupling loss variation over the frequency range used in the measurement which does not exceed 2 dB;

-       a non-linear elements which can affect the accuracy of the measurement results.

The measurement of sensitivity is performed in the following manner;

(a)   Place the equipment in the RFCD;

(b)   Connect an RF signal generator to the input terminals of the RFCD;

(c)   Connect an audio frequency output meter and a distortion factor meter to the audio output of the receiver;

(d)   Adjust the RF output of the signal generator until the squelch just opens;

Note: The squelch sensitivity setting is the same as that used on the radiation test site.

(e)   Note the output of the RF signal generator, this equates to the field strength μV/metre measured on the radiation test site;

(f)   Disable the receiver squelch;

(g)   Modulate the RF signal generator as in the receiver sensitivity test described earlier in the paper;

(h)   Adjust the level of the output of the RF signal generator until a signal-to-noise ratio of 20 dB is obtained at the receiver audio output.

(i)   The increase or decrease in signal level measured in step h) is referred to the level of input signal in step d) to obtain the sensitivity in μV/Metre.

The above measurement of sensitivity can now be conveniently used for other measurements of the receiver, such as adjacent channel selectivity, intermodulation response attenuation, blocking and de-sensitization. The one restriction with the use of an RFCD is the radio frequency bandwidth which is generally restricted to one or two MHz.

11.8 Conclusions

The key to success with radiated radio frequency measurements is attention to detail, such as the instruments used, the test antenna, cables and connectors, and above all, the construction of the test site.

The measurements described only tell part of the story, and have dealt with the equipment under test in a free standing state which, in the main, meets the requirements for mobiles and fixed stations. Equipment with integral antenna, such as radio pagers carried or worn by a person which will effect the equipment performance, in particular the antenna radiation pattern. It is therefore necessary to carry out tests with this type of equipment on a person, which adds another variable to the measurement. Variations of 6 dB or more will be found in practice over a population of people. Work has been carried out in many countries to construct a simulated person, with the objective of creating a world standard, but to-date this has not been achieved!

The measurement described in this paper have not covered all the characteristics of the transmitter and receiver which need to be considered in the overall design specification of the equipment. The measurements described have been restricted to those that are required for type approval purposes which are mandatory for all equipment used in the mobile services. A further purpose has been to highlight the limitations in practical measurement, due to the test arrangements, the performance of test equipment and the limit value chosen for the measurement.

11.9 References

1. SKIFFINS, R.M.: 'Regulatory control of land mobile radio', in HOLBECHE, R.J. (Ed): 'Land mobile radio systems' (Peter Peregrinus, London, 1985)

2. RUDKIN, A.M.: Marconi instruments and measurements Mag, No. 15, June 1977

3. IEC Publication 489, Parts 2 to 6: Methods of measurement for radio equipment used in the mobile services which includes RF radiation test sites

4. DTI, Radiocommunications Agency: MPT specifications for equipment used in the mobile radio services, i.e. MPT 1326 and MPT 1327

5. ETSI: Technical characteristics and test conditions for radio equipment with an internal or external RF connector, intended primarily for analogue speech for use in the land mobile services, ETS-B, 1990

Chapter 12

Radio site engineering

Gerald H David

12.1 Introduction

Mobile radio expanded rapidly in the 1960's and a large number of base stations were established with a high concentration of prime (high) sites, predominantly urban. It could be argued that many of these stations performed a communal function prior to the existence of an adequate engineering specification for such an arrangement; as a result proliferation in the major cities was haphazard, with inevitable interference problems, especially with co-located base stations.

A solution for the operational problems of communal sites requires a system having a single broadband antenna. This solution involves the development of multicouplers for groups of transmitters and receivers at the communal site. The centrepiece of the system are filters which give isolation between adjacent bands and protection from the various forms of interference. These filters represent a significant contribution to present day high efficiency installations, and stem from the concept of controlling the problems at the design stage. The filters have a bandpass configuration, and protect all the equipment connected to the antenna from frequencies outside the relevant band. These filters are appropriately called "Spectrum Dividing Filters", since they divide segments of the spectrum from each other. (Hereafter called SD filters)

A general arrangement for a multicoupler system is shown in Fig. 12.1 and it will be seen that the SD filters are placed in the transmit and receive paths of a multiple system site. Both are coupled to the common antenna by critical cable lengths.

Various combinations of SD filters covering all the frequency bands of interest may be coupled together in a variety of arrangements, to enable single antenna working for the optimum number of base station channels. The high efficiency mentioned above refers to reduced spurious radiations, significant isolation of transmitter and receiver functions and improved signal-to-noise ratio at the receiver. All these advantages are secured during the initial design procedure, using a known degree of isolation between transmitter and receiver, and provision of bandpass filters having the requisite selectivity together with low insertion loss. Clearly, the filter design must also ensure least dissipation of transmitter power and minimum degradation of the receiver noise performance.

The implementation of large scale communal site operation always encourages a wide spectrum of unwanted signals radiated by the various equipment; multicoupled installations alleviate the condition.

In what follows, the details for complete system design with all essential components are set out. The basic module of the SD filter requires complementing with appropriate cavity resonators, a receiver multicoupler system and, in many cases, additional components, depending upon the final parameters required for a given installation, i.e. Fig. 12.1.

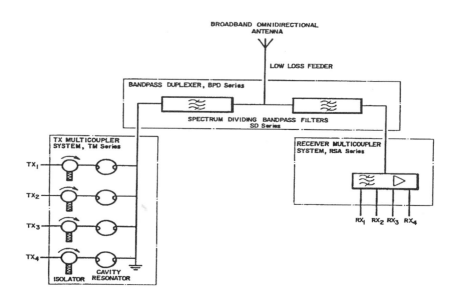

Fig. 12.1   Integrated site block diagram

In the United Kingdom at least, there is considerable pressure on the mobile radio industry from the environmental and aesthetic lobby concerned with preserving the appearance of hilltops and other natural surroundings. This has led to stringent planning laws and extremely close scrutiny of all planning applications for radio structures. The result has been the considerable expansion in the application of multicoupler systems for communal sites, especially those requiring the use of high sites.

The other area in which there is an incentive to employ a multicoupler "Integrated Site" approach, is for the reduction in spurious noise and intermodulation products.

In many cases it is not now possible to work without filtering on London sites, due to broadband interference emanating from adjacent stations. The classic case is the operation of a mobile radio system on sites adjacent to the BBC Crystal Palace mast, where the field strength at VHF and UHF frequencies in the broadcast radio and TV bands often exceeds 10 volts per metre. By the judicious use of bandpass filters, however, it is possible to connect a multitude of mobile radio base stations on the BBC mast itself, and there are currently more than thirty such stations installed at that site.

12.2   The available resources

For any mobile system the starting basis is the establishment of the required coverage (service) area, the location of the base station site and the allocation of a licence. These requirements also usually define the operating frequency.

The most typical situation are the VHF bands as these have in fact been the source of greatest range of experience during the past twenty years. The new Band III, for example, has propagation characteristics with operating parameters closely relating to the original VHF high band, and the upper segment of Band III is contiguous with the military 225-400 MHz spectrum.

At 900 MHz there are new problems emerging, but there is evidence that the control of radiation can be better exercised at these higher frequencies, particularly for cellular type systems.

The filtering techniques described later in the chapter can be applied equally to any of the frequencies under consideration and it is only the physical disposition in accordance with the fundamentals involved that need to be changed.

There is a current tendency for the coverage requirements from a given site to be diminished with the increased use of cellular type systems, and with further developments into high UHF and possibly microwave frequencies for mobile radio, the planned coverage distances may reduce even further.

A typical requirement at 900 MHz for example, is a range not exceeding 5 km and in the case of proposals above 1 GHz, there are likely to be practical systems having a coverage range of less than 500 meters. Such systems are only practical when they are linked together to form coordinated structures of coverage, and the system efficiency then relies heavily on the infrastructure.

Most of the traditional private mobile radio systems in current use have a typical range of 15 - 20 miles but the bulk of these systems are not interconnected to provide continuous coverage through a complex infrastructure. All of the above types of systems and the solutions that emerge can be applied to current proposals.

A further problem within the mobile radio systems is the co-channel interference caused, often by misuse of the system or by excessive power being applied to situations where adequate coverage can be obtained with a much lower transmit power. It is often not realised that in the UK we have a much higher density of mobile radio per given area than almost any other country in the world, and the experiences obtained elsewhere, particularly in the USA for example are based on an entirely different distribution and utilisation schemes.

12.3   System parameters and design objectives

The radio equipment for the mobile and fixed service is built to specifications which are directed to ensure the efficient use of the radio spectrum. One set of parameters control emitted bandwidth and the level of out-of-band radiation which will cause interference to other users; they establish suitable transmitter power or effective radiated power (ERP) limits and will specify the receiver sensitivity and limits to the levels of spurious emission from receivers. Another set of parameters define conditions which make a system less susceptible to interference by others; they include receiver selectivity, dynamic range and blocking characteristics.

Good installation design ensures that as far as possible the performance of a complete installation preserves the professional characteristics of the components, laying down the intended field strength in the designated area, avoiding the radiation of spurious emissions and preserving the sensitivity of receivers.

The objectives are as follows:

(a)   to obtain the coverage required from the chosen site in a precise and well defined manner;

(b)   to cause minimal spectrum pollution to other users on adjacent sites;

(c)   to cause minimal interference to other co-sited users;

(d)   to operate the system with the ERP and optimum spectral efficiency compatible with providing the required service;

To fulfill the requirements of all relevant legislation and recommendations, the above criteria should be met for the *whole of the working life of the installation*, and should allow for future expansion.

Typical parameters for a private mobile radio installation are as follows:

BASE STATION

(a)   Transmit power 25 W (typical) CW (+44 dBm)

(b)   Antenna system unity gain, omnidirectional

(c)   Channel bandwidth 12.5 kHz

(d)   Service range 15 - 17 miles

(e)   Receiver sensitivity 0.3 μV emf (-117 dBm to threshold)

MOBILE EQUIPMENT

(a)   Transmit power 10 W (typical) CW (+40 dBm)

(b)   Antenna system unity gain, omnidirectional

(c)   Channel bandwidth 12.5 kHz

(d)   Service range 15 - 17 miles

(e)   Receiver sensitivity 0.5 μV emf (-110 dB to threshold)

It can be seen, therefore, that a dynamic range exceeding 150 dB exists between the base station and the mobile, of which approximately 6 dB will be taken up by losses in the radiating system at both ends, i.e. feeder loss, etc., leaving more than 140 dB as the working propagation loss for the system.

This means that for an unobstructed path, i.e. a high site overlooking a wide, flat plane, the signal may be workable over a range not even limited by the earth's curvature. This can give rise to excessive propagation, i.e. upwards of 40 miles, and as a result co-channel interference from adjacent systems.

In conurbations where the antenna height may be limited and the propagation path obstructed by a variety of man made structures, the radio range will be severely reduced, and a typical working range of 12 miles is considered reasonable, say, in the central London area.

12.4   Typical problems on complex sites

12.4.1   Site planning

It can be generally stated that all good radio sites are on hilltops which have visual attraction and form part of the skyline. It is usual, therefore, that all new proposals for masts and towers meet with considerable hostility and opposition from residents, planning authorities and environmental bodies throughout the country often as the result of previous insensitive approaches from statutory bodies in requiring large structures.

It could be fairly stated therefore that the land mobile industry is generally beset with stumbling blocks when it comes to expansion of base station sites, masts, towers, antenna and the means generally of propagating signals.

The unpopularity of these structures is not at all helped by telephone, power cables and other masts and structures which also dominate the landscape, and it could be generally concluded that no new structures can be erected without an outcry of one sort or the other. For example there are now moves afoot to restrict mast development in the Home Counties and the problem has taken on even greater dimensions, in view of recent proposals for large scale development in the land mobile radio telephony.

12.4.2   Channel allocation

It is highly likely that channels allocated for mobile systems will be shared in certain areas, although attempts will have been made to have wherever possible a nationally dedicated channel. There will therefore be co-channel users and all their associated difficulties.

12.4.3   System deficiencies

There are many factors which cause the system to be less efficient than the required parameters outlined above, and the most significant is unwanted signals appearing in the receiver. These can arise from co-channel users in adjacent areas, which can only be resolved by mutual arrangements of antennas, radiated power, etc. The most common problem, however, is identified with signals generated in and around the site.

12.4.4   On site unwanted products

There are three main sources of radiated products:

(a)   the noise and spurious products generated within transmitters; these occupy a broad bandwidth on both sides of the carrier frequency.

(b)   intermodulation products caused by the mixing of two or more source frequencies which produce well defined and often high level signals. These are normally caused by transmitters coupling into an adjacent transmitter output stage, due to inadequate isolation between the two transmitters.

(c)   intermodulation products caused by non-linear effects on the mast and antenna hardware.

(d)   intermodulation, cross modulation and blocking effects in receiver systems. Problems are usually caused by large signals at the input of the receiving system causing non-linearity.

The specifications for receivers are well defined in existing documents; distribution amplifiers may be called upon to operate in a more hostile environment on densely utilised sites and require a mandatory specification.

12.5   Mechanisms of interfering sources

1.   Simple frequency relationship

As land mobile frequency bands are allocated throughout the VHF/UHF spectrum there may be harmonic relationships between frequencies in the various bands. The equipment cabinet, the power supply cabling and land-line cabling can contribute to the level of these unwanted harmonic signals.

Other interfering signals can be caused by simple mixes either in transmitter output stages or at the antenna mast. As an example, if the signal from a VHF broadcasting transmitter at 93 MHz mixes with a signal of the mobile service at 170.5 MHz, a difference signal of 77.5 MHz can be produced. This can cause a problem if it is a receive frequency of another mobile service.

2.   Complex frequency relationships

 (i) Generation of the intermediate frequency and/or its derivatives.

       Interference can be caused in a receiver where signals are received from
       two transmitters whose frequencies are separated by an amount equal to
       the IF, or a submultiple of the IF, of the receiver.

 (ii) Generation of transmit/receive (Tx/Rx) difference frequency

       This problem arises on sites where there are several base stations having
       "repeater" or "talk through" facilities, i.e. the transmitters and receivers
       are in use simultaneously. If the Tx/Rx spacing is constant, an incoming
       signal from a mobile station will produce in the base station transmitter
       output stage a difference frequency. Any other base station transmitter
       may now mix with to produce its own receiver frequency in the same band.

3.   Intermodulation products

 (i) Generated external to the site.

       Under this heading, products arise from stations on adjacent sites and, in
       particular the third order product, i.e. $2f_1 - f_2$, which is prevalent in large
       built-up areas. In some instances significant intermodulation products up
       to and including the seventh order have been noted, and in exceptional
       cases the interference has been traced to the nineteenth order.

 (ii) Intermodulation products generated on-site by non-linear junctions on the
       mast.

       More study is required to verify the mechanisms and levels of such
       interference, which certainly exists in the land mobile bands. However, at
       lower radiated powers, the significance of these products is reduced,
       compared with other forms of non-linearity.

(iii) Intermodulation products generated on-site by non-linearity in compo-
       nents of the system.

       Junctions between dissimilar metals cause non-linearity, and therefore
       intermodulation products, when subjected to radio frequency currents,
       and recent work has highlighted such products up to the eleventh order at
       VHF caused by connectors, cables and dissimilar junctions in what might
       be regarded as otherwise innocuous components.

       For the long-term development of the land mobile radio industry, it may
       be necessarily to define the non-linearity of passive components in the
       system.

4.   Transmitter noise

Until quite recently, most transmitters at base station sites had valve output stages, which remarkably are not a major contributor to the noise spectrum compared with the more modern solid-state output stages.

With a valve output stage, the unwanted noise is generally narrow-band, having frequencies which are multiples of the crystal oscillator frequency or a combination derived from the multipliers. However, in the case of solid-state output stages the noise is generally wideband and higher in level.

Figs. 12.2 and 12.3 give the results of measurements made of noise from VHF transmitters with a thermionic and solid-state "high band" (150 - 170 MHz) transmitter, respectively.

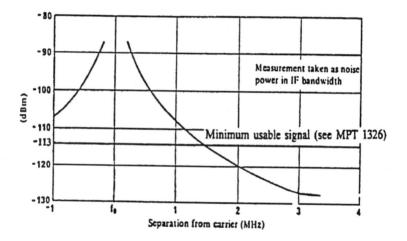

Fig. 12.2 Typical transmitter noise, VHF high band,
          thermionic output stage

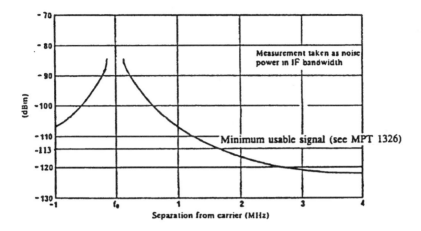

Fig. 12.3 Typical transmitter noise, VHF high band,
solid-state output stage

5. External electrical noise

Apart from ignition noise, there are the several sources of radio interference,
which continue to proliferate, particularly from industrial users, e.g. RF heating,
microwave ovens, X-ray and medical equipments. These normally provide a
broad spectrum of noise which tends to vary in frequency.

Screening or suppression of the interfering equipment normally reduces the
problem to an acceptable level.

There is, however, a new family of sources, namely computers and computer
peripherals, which are currently causing problems with broadband noise over the
VHF spectrum.

12.6 The integrated site concept

Having considered the difficulties that are listed in the previous section, it is
apparent that the radio site must be considered as a single entity.

In the past, each system was brought onto the site as and when the need arose,
often without consideration of existing installations, and typically having an
additional antenna adjacent to an existing user. From this dilemma the concept
emerged of a site with one antenna and feeder per frequency band, and with each
antenna strategically placed to obtain maximum isolation from its neighbours.
Each additional user would have the system planned in advance of being
connected, so that an overall frequency computation could be carried out to
identify intermodulation products. Such a site was called the "Integrated Site",
since it was intended to have a minimum number of antennas with a maximum
number of transmitters and receivers, with all parameters fully controlled.

This concept calls for a set of multicoupled components, which by definition would fulfill the role previously undertaken by separate antennas and feeder cables. There was therefore immediately identified a definition of insertion loss, isolation and other parameters to enable the system performance to be equal to, if not better than, the previous arrangements. The cost of undertaking this work had also to be commensurate with the original cost of a folded dipole, 70 metres of UR67, with no filters involved.

If the coupling arrangement had cost, say £200 more than the originally envisaged arrangement,it would probably have been unacceptable, since it took away profit from the radio site operator. The advantages that were to be gained were mainly in terms of system noise and reduction in interference levels, which can be immediately interpreted as improvement in system coverage. There were some difficulties also in providing antenna systems with sufficient bandwidth to cater for the multiplicity of channels connected in the integrated scheme, as it was now necessary to consider bandwidths of the order of 20% of centre frequency.

It has not been possible to find existing antennas with precisely the parameters needed to fulfill the role outlined by the integrated site, but a commercial compromise has been reached whereby the antennas are of a performance and quality that give sufficient bandwidth to allow a significant reduction in the number of antennas, whilst giving coverage of the working area.

The most important step forward is the concept that ten transmitters and ten receivers on a site does not necessarily mean twenty antennas. The first diagram illustrated the basic concept, and identify the modules that constitute the integrated site system.

12.6.1   The necessary modules

The common antenna working concept is dependent on a single feeder cable from an antenna having the bandwidth necessary to cover that portion of the spectrum which has been allocated to the transmit and receive segments of the appropriate band.

Since this feeder replaces a multiplicity of separate feeders and antenna systems it is economic to employ a large, low-loss conductor such as Kabel metal CF 7/8 in. CU2Y. This cable has an overall diameter of 28 mm and an insertion loss of 1.16 dB per 100 metres at 100 MHz. (This compares very favourably with UR67, which has an insertion loss of 7 dB per 100 metres at 100 MHz).

At the equipment room termination of the feeder cable, a pair of spectrum dividing filters are coupled together by means of critical quarterwave harness sections to the antenna feeder. One of these bandpass filters covers the transmit range of frequencies, and the other is for the range of the receiver frequencies, i.e. at VHF "highband" the Tx frequencies extend from 165.025 MHz - 168.225 MHz, and the receive frequencies from 169.6875 - 173.225 MHz respectively.

A common practice for operating duplex systems on a single antenna is by means of bandstop filtering, and the typical duplexer performance is geared to single channel working, i.e. the bandstop filters in each path were tuned to reject a single, narrow band of frequencies. The parameters to obtain adequate duplex working, i.e. on the transmitter and receiver operating simultaneously were as follows:

Tx-Rx isolation: 60 dB min
Tx insertion loss: 2 dB max
Rx insertion loss: 2 dB max
Isolation obtained in the
Tx path at the receiver frequency:
          thermionic circuits   60 dB min
          semi-conductor   75 dB min

These parameters must be taken in conjunction with the typical folded dipole and UR67 installation, which gave a typical overall transmission loss of 4 dB.

Taking into account the integrated site philosophy, the feeder loss could be reduced by perhaps 2 dB, and the antenna gain increased by perhaps 3 dB, thus giving an overall 4 - 5 dB improvement over the original set-up.

It was thus conceivable that the losses in a multicoupler system could equate to 3 - 4 dB, without the overall system performance degrading below that of its predecessor.

It is now evident that complete control of spurious and intermodulation products is possible, the basic modules being as follows:

1.   A means of coupling the transmitters together either by cavity resonators of a high Q coupled together by critical harness, or by hybrid couplers with their inherent loss, or a combination of the two.

2.   The receivers could be coupled together by means of a bandpass filter, followed by an amplifier to make up the losses of a passive dividing network, which would be followed by a Wilkinson type hybrid divider [1].

3.   As a guarantee against intermodulation products being generated in the transmitter output stage, a directional property was required, and this is provided by ferrite isolators. These would normally be placed immediately after the transmitter and before the cavity resonator.

These are the building blocks from which the system can be put together and were shown in Fig. 12.1.

12.6.2   The necessary parameters

Guideline values are shown here:

TRANSMITTER PATH

Inserting loss, feeder cable:1-2 dB max
Insertion loss, SD filter:1.5 dB max
Insertion loss, cavity resonator:1.5 dB max

Insertion loss, ferrite isolator:1.2 dB max
Tx-Tx isolation:40 dB min
Tx-Rx isolation:75 dB min

RECEIVER PATH

Gain of distributed amplifier:15 dB min
Insertion loss of distribution: network:10 dB max
Overall gain of Rx path:2 dB min

Referring back to the building blocks described above, it is possible to envisage a system where the overall characteristics have considerable technical advantages over the original and the service provided to the end user is improved in terms of system noise and system availability.

It is important to note that an insertion loss of less than 1.0 dB has to be achieved in each and every one of the components described above. Therefore, an insertion loss in the range 0.5 - 0.7 dB per component of the system is necessary, to achieve this objective.

12.7 Multicoupling equipment

There have been a wide variety of designs produced over many years, largely as a result of environments demanding a single antenna solution, where the cost or indeed any economic considerations were over-ruled by considerations such as mast loading, security, i.e. the hidden antennas, or aesthetic objections to additional antennas.

In many cases the system comprised simply a Wilkinson [1] hybrid, constructed from cable sections, and designed to give a minimum of 20 dB isolation between the transmitters. The loss of definition would be 3 dB plus the insertion loss of the line section, which would be typically 0.3 dB.

The other systems in common use comprised cavity resonators between 3 in. and 6 in. in diameter and nominally of quarter wavelength construction, giving unloaded Qs in the range 2000 - 4000 in the 150 MHz band. The losses of such systems always amounted to 3 dB or 4 dB and sometimes exceeds 6 dB.

The first attempt at low loss multicouples was undertaken as a result of work done by RAE Farnborough for ground-to-air communications, where the need for additional channels was increasing dramatically, and the loading on masts at ground stations had reached unacceptable levels.

A quarterwave coaxial cavity resonator was designed and manufactured in large quantities between 1964 and 1970 having an unloaded Q of 2500 at 136 MHz, an insertion loss of 2.5 dB and rack-and-pinion type tuning which adjusted the length of the centre conductor assembly. The construction was in brass and copper, silver plated throughout, and the cost for the first time came below £100 per unit in quantities exceeding one hundred.

At about the same time the BBC, [2], produced a two and three port combiner for UHF television frequencies to enable the IBA to share BBC sites, and vice-versa for UHF transmissions. The BBC combiners were also based on quarter-wave resonators, but since the powers handled had to exceed 40 kilowatts and the quantities were numerically quite small, cost was not such a major consideration, and the resulting design more closely approached the performance derived from the theoretical fundamentals of quarterwave resonators than any units previously manufactured in industry.

Weight and cost also became very apparent as major considerations for the long term development of the multicoupling techniques, since brass and copper, the favoured materials, were neither cheap nor light when compared to, say, aluminium or plastics, which are favoured nowadays for many engineering applications.

Another aspect of the fundamental design was signal leakage from the component into adjacent units, or, worse, from adjacent transmitters into the component. Considerable attention therefore was given to sealing of joints and faces, particularly where a large expanse of metal plate was used to close a multi-section helical line resonator, where the out-of-band attenuation on the critical frequencies was expected to be in excess of 90 dB.

A very commonly overlooked aspect of all multicoupling systems is that despite having a filter system with extraordinarily good performance, e.g. greater than 90 dB isolation, all this can be thrown away by coupling leakage external to the components or the system. A very good example of this is the use of inferior feeder cables bundled together on one cable tray, where for example the transmitter cable and the receiver cable of a combiner system could produce coupling below 60 dB, when the filter network had already given the system operator an isolation of, say 90 dB.

12.7.1   Components of the system

(i)   Spectrum dividing filter

The spectrum dividing filter is the title given to a bandpass filter designed specifically to cover the bandwidth of a defined segment of the mobile radio spectrum, and having a very sharp cut-off at the band edges where, for example, in VHF "highband", the Tx/Rx spacing is 4.8 MHz, i.e. Fig. 12.4.

A filter was required, therefore, that would pass the associated groups of frequencies, while preserving an isolation at band edges not less than the minimum to enable degradation free operation of all the receivers when the associated transmitters were keyed.

With thermionic equipment it had been possible to work in duplex mode with as little as 55 dB of isolation between Tx and Rx, but with the advent of solid state systems the general increase in spectrum noise, largely due to semi-conductor junction noise, made it necessary to increase the isolation to at least 65 dB, and in many cases 80 dB.

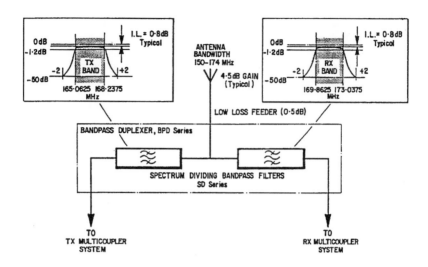

Fig. 12.4   Spectrum dividing filters in a multicoupling system

The requirement for the filter, therefore, is that at 168.500 MHz, the insertion loss is still acceptably low, normally defined as 1.5 dB, whereas the isolation at 169.8 MHz has to be at least 65 dB. After much experimentation, employing various combinations of single cavities and multiple section filters, the best compromise for such a helical filter was found to be a Butterworth type of response suitably modified according to Zverev [3] to incorporate a degree of Chebyshev outband characteristics, without introducing too much inband ripple.

It was found that a pure Butterworth design had the deficiencies that, although the in-band response could be made flat and held within ripple limits of 0.1 dB, the skirt attenuation was insufficient to enable to overlap attenuation characteristics to be obtained.

In any case the Zverev filter is very well defined and gave practical realisation of the best compromise then available. Subsequent work carried out has modified the designs in order to obtain the very best practical compromise between low insertion loss and outband attenuation. Typical responses are shown in Fig. 12.5.

The filters can be generally described as helical resonator, multi-section, slot and aperture coupled, having a Butterworth-like response. In the band under consideration attention was given to both maximum isolation as close to the passband as possible. In order obtain maximum unloaded Q the helical resonators consist of approximately three and a half turns of copper tube, supported adequately by p.t.f.e. posts. The sections each comprise a resonator, separated by an aluminium block dividing the space between the sections, forming a bulkhead which is rigidly fixed and having the upper segment cut off to reveal the adjacent section. Fine adjustment is achieved by a set screw which is fitted on the centre line of the partition or bulkhead. The frequency of resonance of the helices is adjusted by means of a tuning screw which is inserted centrally in the open end of the helix (i.e. at the end away from the earthed end).

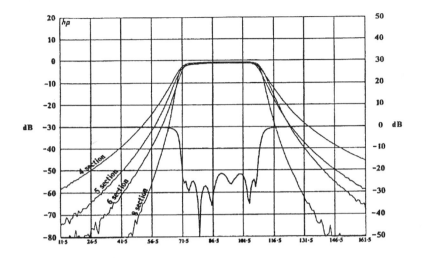

Fig. 12.5 Composite response of an SD filter

(ii) Cavity resonators

As a start, a traditional American industry standard, the 6 in. aluminium high pressure gas pipe, cut off into approximately quarterwave long sections with end plates welded to the extremeties and having an adjustable centre conductor to tune a bandwidth of ± 5% of centre frequency. A wide variety of companies made such a device, and one company used a copper equivalent. The Q_0 realised by such structures, however, never exceeded 4500, and when loaded to practical limits would give an operational Q of 700 and an insertion loss of approximately 3 dB.

On examination of the outband resonance curve the device has an asymptotic isolation of only 30 dB; with care 33 dB may be achieved.

The isolation at 0.1% of centre frequency would be only 4 dB, and therefore as a coupling structure for the UK highband - for example in central London - frequency spacings are such that Tx/Tx separation of only 100kHz became essential.

Experiments were carried out to establish the parameters of a resonator to give a budget of 15 dB isolation at 150kHz Tx/Tx spacing, and with an insertion loss of 1.0 dB.

This gave a requirement for an unloaded Q greater than 10,000, a loaded Q of not less than 1000 and a physical size of approximately 2.5 cubic feet based on known structures, which were physically measured and interpolated to the required dimensions [4].

Various structures were built in aluminium, copper and brass, both in square, rectangular and circular section, to establish a compromise between performance and economics. It soon became apparent that the structure of the cavity had to be seam-free, or else welded, and without internal surface blemishes, in order to produce a reasonable effective Q when compared to volume, as even a very large cavity with a poor surface conductivity did not give acceptable unloaded Q.

The best results were obtained with a rolled copper structure, silver plated and with an outer casing of brass brazed at all seams, and integrally linked by brazing with the inner copper structure. With such a system it was possible to come within 90% of theoretical maximum Q available from the volume of the structure. In terms of economics, however, this arrangement was prohibitively costly, and alternatives were sought that already existed in some suitable form as conductive, smooth, mass-produced units.

The chemical industry and all the process industries were examined for pressure vessels, e.g. sterilisers, steam baths, degreasing plant etc., and the most suitable ready made unit came from the brewing industry, namely, an eleven gallon I.E.C. aluminium beer barrel.

This structure had the necessary volume, a smooth inner surface, contiguously welded seams throughout, and was made from high conductivity aluminium. Production was readily available in thousands, and the price was far lower than any alternative process to create such a structure. The electrical performance of one such resonator for VHF low band (hence the necessity large volume) is shown in Fig. 12.6.

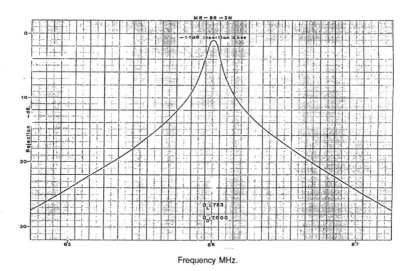

Frequency MHz.

Fig. 12.6   Cavity resonator - typical response

(iii) Receiver splitter amplifiers

A fundamental deficiency in many early mobile radio systems was the disparity between the transmitter power radiated form the base station when compared to the mobile equipment which was usually of a much lower power, had a very poor feeder system, and often a very inefficient antenna system.

In many cases the disparity exceeded 10 dB, and therefore a common deficiency existed whereby the mobile receiver could hear the base station transmitter well beyond the point where the mobile transmitter was unable to be heard by the base station receiver.

The writer examined the possibilities of remedying the situation by equalising the power loss in both directions, and without modifying the mobile installation, and the simplest was to provide a high gain, low noise amplifier at the base station receiver.

An examination of the amplifiers available revealed that semi-conductor devices existed for small signal amplifiers having claimed noise figures of less than 3 dB, and several trial designs were built using these devises.

The good device was a field effect transistor having a claimed noise figure of 2 dB when used in the test circuit configuration offered by the manufacturer. It was soon found, however, that when used in a conjugate match configuration it was virtually impossible to obtain a final noise figure below 3.7 dB at frequencies for example between 450 - 470 MHz. All of the parameters were varied, e.g. the d.c. conditions, the method of matching and the type of assembly in order to obtain a record of the possible combinations to achieve (a) optimum noise figure (b) maximum gain and (c) optimum match conditions.

A typical acceptable final result was as follows:

Gain:	Not less than 16 dB
Noise figure:	Not greater than 4 dB
Return loss at the Input end:	Not worse than 18 dB
Return loss at the Output end:	Not greater than 16 dB

The installation arrangement is shown in Fig. 12.7.

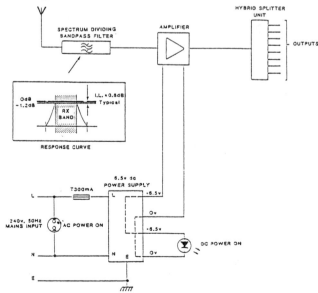

Fig. 12.7 Receiver distribution amplifier system

(iv)   Ferrite isolators

A vital component in the protection of the transmitter is the unidirectional characteristic of the ferrite isolator. The port-to-port isolation provided varies in accordance with the bandwidth, which is a function of the way in which the device is fed and loaded, i.e. by the use of series capacitance in the ports a very narrow bandwidth may be achieved by resonating the natural characteristics of the device in a series mode.

When the third port is terminated in a suitable dummy load, the isolator becomes a circulator, and while giving a through insertion loss of typically less than 1.0 dB, the reverse characteristic is usually better than 25 dB, and often exceeds 35 dB, i.e. Fig. 12.8.

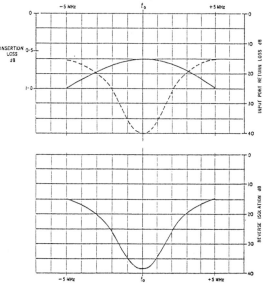

Fig. 12.8   VHF isolator performance characteristics

The position of the isolator with respect to the cavity resonator and SD filter was indicated in Fig. 12.1.

(v)   Hybrid combiners

The experience gained from the work done in dividing networks for the receiver distribution amplifier was built upon for higher power devices, again using the basic Wilkinson concept.

A wide range of excellent quality 75 ohm and 82 ohm cables exists, from which it has been possible to build a wide variety of 3 dB couplers, hybrids and phased line arrangements, combined together as shown in Fig. 12.9.

The bulk of the components are aimed to cover the range 25 - 50 W input power, and are four port, the diagonal ports being termed as 'input connections' and the two remaining being 'output' and 'load' parts respectively. It is normally possible with accurate cutting to within 0.2 mm that the natural balance without compensation across the hybrid exceeds 35 dB, and will often be better than 45 dB.

The use of variable capacitance to adjust the balance has been avoided, as this becomes a critical frequency adjustment requiring constant attention and immediately negates the "fit-and-forget" philosophy.

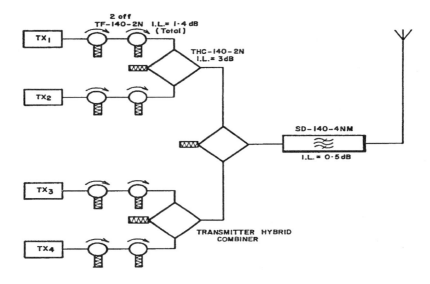

Fig. 12.9   Adjacent channel TX system (12.5 kHz channel separation)

12.8 Practical system examples

There are now many typical examples of working systems where a combination of bandpass filters and other components have been put together in order to solve a particular site engineering problem.

In many cases these designs have enabled systems to be put into service that otherwise without this technology would not have been possible.

It should be noted that in many cases more than one antenna have had to be employed due to the close proximity of transmitter frequencies.   This is a compromise between insertion loss and acceptable mast loading, not forgetting, or course, the cost of the system.

In every case, however, a non filter coupled proposal would have meant that the existing mast could not carry the additional loading.   In each case therefore there has been a considerable saving, not only of mast costs, but disruption to the existing services while alterations took place to the site.

The addition of another mast on existing sites is also likely to cause propagation path attenuation, and even in some cases total disruption to existing services.

Fig. 12.10 shows a typical arrangement for 900 MHz operation. The electrical requirements are as stringent because even though the Rx/Tx frequency separation is 45 MHz, the frequency is now nearly nine times larger compared to VHF operation.

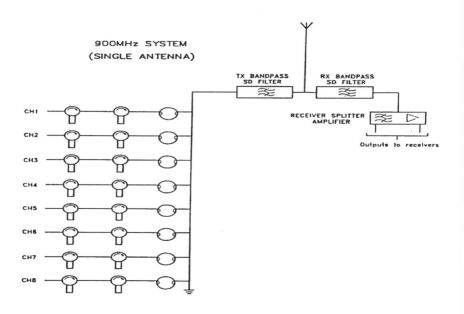

Fig. 12.10   900 MHz system (single antenna)

12.9 Conclusions

It has been shown that with the increased use of mobile radio the problems encountered on communal sites can severely inhibit the design of systems unless a well filtered compromise can be found.

The final performance of the system is largely dependent on the ease with which mobiles at the extreme edges of the coverage plan are able to communicate with the base station at an acceptable signal-to-noise level.

The use of filters has greatly increased the opportunities for multiple use on complex sites and future systems are likely to depend heavily on this type of solution.

12.10 References

1. MANTON, R.G.: 'Hybrid networks and their uses in radio frequency circuits', The Radio and Electronic Engineer, Volume 54, 11/12, pp. 473-489, November/December 1984

2. SHONE, A.B., and WHARTON, W.: 'Transmitter combining networks for television transmitting stations', International Television Engineering Conference, IEE Conference 1962, Series No. 5, Paper No. 3950E, pp. 462-472

3. ZVEREV, A. I.: 'Handbook of filter synthesis' (John Wiley & Sons, Inc., 1967)

4. HOWSON, D.P., and AL HALFID, H.T.: 'The design of cavity resonators for VHF mobile radio base stations', IERE Conference on Civil Land Mobile Radio, Proceedings No. 33, November 1975, p. 57

Chapter 13

The pan-European cellular mobile radio system

Professor David Cheeseman

13.1 Introduction

The aim of this chapter is to provide a general description of the pan-European cellular radio system, commonly known as the GSM system, detailing in particular the structure, operation and facilities, but not the technology or implementation. The chapter also provides some of the political and industrial, as well as the technical, background which led to the formulation of the GSM system standard. The development of the standard has been strongly influenced by political considerations; however as this is a technical chapter a detailed history of events would not be appropriate, but a few examples are given. The basic principles of cellular radio are assumed as described in Chapter 9.

Here an attempt is made to answer three questions about the GSM system;

- What is it?

- What are its features?

- How does it work?

The answer to the first question - What is it? - is quite simply;

The GSM pan-European Cellular Mobile Radio System is a

- Public Telecommunications System

It is not just a radio system, which will become clear as the various requirements are explained. Whereas there are many mobile radio systems that are not connected to public telephone networks, those that are must conform to public telecommunications system standards and possess recognisable features. For example, it must be possible to make and receive telephone calls in the same general way as in fixed networks; users must be able to dial up numbers on the same and other fixed and mobile networks, in the local area, and via national and international direct dialling using broadly the same procedures as for fixed network calls. Similarly, incoming calls from anywhere in the world must be possible. Data and messaging services must also be available; in short it must behave like a telephone ... without wires.

A public telecommunications system must offer services and facilities that are contained in a continuously growing internationally agreed schedule of speech and data bearer and teleservices, although the constraints imposed by the mobile radio link result in some restrictions in the range and the quality of certain facilities in mobile networks.

13.2 Background

It was with this highly intricate web of telecommunications standards in mind that the pan-European Cellular Mobile Radio System was conceived in 1982 by a committee of the Conference of European Posts and Telecommunications Administrations (CEPT)[1]. CEPT foresaw a growing need during the 1980s and 1990s for public cellular radio, but accompanied by a widening divergence of systems, unless a standard was available to positively encourage convergence. It also wished to achieve a measure of compatibility with the ISDN. The system that has emerged from this process has become widely known as the GSM System.*

An outline of the structure of CEPT at the time is shown in Fig. 13.1 (official titles are in French as shown here).

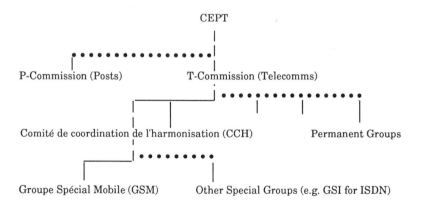

Fig. 13.1 Outline structure of CEPT in 1982

---

*   GSM now stands for 'Global System for Mobile Communications'; previously it was named after the planning group described in para 13.3.

Early in 1989 GSM was transferred from CEPT to the European Telecommunications Standards Institute (ETSI), but its programme of work continues as planned. CEPT continues to provide a forum for European telecommunications standardisation, for example on call charging arrangements and radio frequency allocations which, however, are outside the scope of this chapter.

ETSI is a legal entity, has over 215 members among the industry community as well as the telecommunications operators, and a permanent staff of about 36 located in purpose-built accommodation at Sophia Antipolis, near Nice, in the Alpes Maritime, France. It deals with a wide range of telecommunication standardisation matters through a General Assembly and a Technical Assembly which meet several times a year. The TC/GSM is currently the largest Technical Committee; the relevant part of the structure of ETSI is shown in Fig. 13.2.

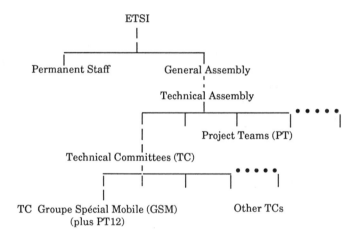

Fig. 13.2   Outline structure of ETSI in 1989

In 1981 a coordinated cellular system, the Nordic Mobile Telephone System (NMT450) using analogue technology and operating in the 450 MHz band, was opened in the four Nordic countries Denmark, Finland, Norway and Sweden. Elsewhere in Europe, the USA and Japan research was fairly well advanced but public service had not commenced; indeed the original papers describing the ideas had been published decades earlier and non-cellular mobile radio systems had already been in use since the 1920s. However, it was not until the emergence of reliable microelectronic circuits and particularly microprocessors in the 1970s that practical cellular systems could be realised [2].

By 1982 various European administrations were showing interest in cellular mobile radio; the market was growing rapidly and the Nordic experience was soon to prove prophetic. Growth always outstripped forecasts, and it continues to do so still. In several countries industry was gearing itself to supply analogue cellular systems to any operator anywhere in the world. Britain was seeking a system for introduction in 1985 although, sadly, UK industry showed little interest in supplying what was still regarded as a specialist market.

13.3   Aims of GSM

It was against this background that the Groupe Spéciale Mobile (GSM) was set up in 1982 specifically to develop a pan-European public mobile radio standard and given the following tasks:

- December 1986 - Outline Definition

- December 1988 - Specification of a pan-European Public Land Mobile Network (PLMN)

The object was to create the possibility of opening public service throughout Europe in the early 1990s. At the outset it was recognised that the best way of achieving this was to create a new European-wide PLMN standard. The intended outcome was, however, visualised only in general terms and the full extent of the standard did not become apparent until a fairly advanced stage in its development.

The aims of the GSM system are to provide:-

- European-wide roaming and

- a mass market for low cost cellphones

In 1982 the target dates looked easily attainable but, as the true nature of the task became progressively clearer, they began to look quite unrealistically early. However, the effort was increased and more sharply focused on the different activities so that those milestones were broadly met and, therefore, service is indeed expected to open in 1991.

The way was paved by a number of bold technical decisions, sometimes ahead of the necessary confirmatory research and development work, or based on simulations and trials carried out under the supervision of a GSM WP. One of the most technically far-reaching decisions, made in 1987, rested on the results of radio and speech coding trials held in Paris, which pointed strongly towards the adoption of digital techniques. Thus the GSM system became a digital system.

By early 1983 several competing system proposals had emerged. Each of the members of GSM had strong preferences and it soon became clear that for various reasons there was no single existing system proposal acceptable to all members as a candidate for the pan-European system. Therefore one of the first tasks of GSM was to find a political formula that would facilitate progress. This formula defined all the existing systems and system proposals as "interim" in order to permit their introduction into service in various countries to cater for the burgeoning demand without prejudicing the ultimate development of a new, truly pan-European, system.

This had two results, one was to steer the thinking of GSM away from current developments, and the other was to set it on the path of political involvement which was to become more significant year by year.

13.3.1   Availability of spectrum

By 1983 several countries were operating systems in the 450MHz band and growth in the Nordic countries was already generating pressure for a move up to 900 MHz. Spectrum in this band was already assigned by the International Telecommunication Union (ITU) for public mobile use in Europe through the mechanism of the World Administrative Radio Conference (WARC) but not yet allocated by all the individual administrations of CEPT.

The ITU divides the world into three zones for spectrum assignment. In Europe the 900 MHz PLMN bands occupy 2 x 25 MHz; the uplink (mobile station transmit, base station receive) band is from 890-915 MHz and the downlink (base station transmit, mobile station receive) band is from 935-960 MHz; the separation between the uplink and downlink bands, the duplex separation, is 45 MHz.

The UK was the first country to allocate part of this band for its first generation, analogue, system - Total Access Communications Systems (TACS) [3] - and it was important for CEPT to reserve some spectrum for the GSM system and to defend it against the pressure of growth while the pan-European standard was under development.

Therefore another of the early duties of GSM was to establish that radio spectrum would be available throughout the whole of Europe; the block thereby reserved for the GSM system is the top 2 x 10 MHz. Thus while the early or interim systems would progressively occupy the lower 2 x 15 MHz, the top 2 x 10 MHz would be protected - protection that proved vitally necessary as the interim systems grew with rapidly increasing demand for service.

The 900 MHz band is now widely used throughout Europe and due to the phenomenal growth of cellular radio in the UK additional spectrum just below the original allocation has been made available, at present in a limited geographical area but soon to be nationwide (Extended TACS or ETACS). Thus when the GSM system is introduced it will face the existing analogue systems occupying the lion's share of the available spectrum, thereby emphasising the need to achieve high spectral efficiency. In time, as the GSM system grows, it will migrate into the lower part of this band displacing the analogue systems and eventually occupying the whole 2 x 25 MHz and, no doubt, extending into additional spectrum as TACS has already done in the UK.

13.3.2   Organisation of GSM

Originally the terms of reference of GSM required it to coordinate the standardisation work of other relevant CEPT groups, but not to initiate new work. GSM was expected to meet about 3-4 times a year to discuss written contributions; this soon proved to be unrealistic.

Various pressures, from operators and Administrations, from the market place, from industry, and from the European Commission, have brought about changes in the intervening years and eventually GSM was authorised to originate as well as to coordinate specific studies. A Permanent Nucleus (PN) based in Paris and specialist Working Parties (WP) and Experts' groups (EG) were set up to support the work.

The PN has been permanently staffed by personnel on secondment from different administrations. However, the WPs and EGs have relied on the willingness of employers and employees alike to engage in a heavy schedule of preparation of contributions of great technical depth which they then have presented, argued, defended, amended (or even withdrawn) after detailed discussion at the typically week-long meetings held sometimes at four to six week intervals throughout the year.

The structure and chairmanships of GSM just before its transfer to ETSI are shown in Fig. 13.3, while Fig. 13.4 shows the current (1990) organisation. The strong involvement in relation to their size of the Nordic countries can clearly be seen and this reflects the importance to these countries of mobile communications. These charts do not, of course, tell the whole story and perhaps hide the contribution made by many other countries.

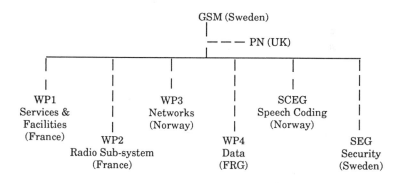

Fig. 13.3 Structure and chairmanships of GSM under CEPT

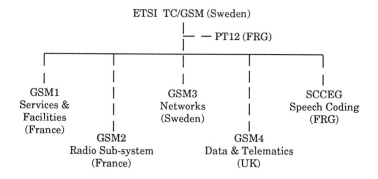

Fig. 13.4 Structure and chairmanships of GSM under ETSI

Under the auspices of another European body, European Cooperation in the Field of Science and Technology (COST), which was in fact jointly responsible for the speech coding experts' group (SCEG), a new group, the speech and channel coding experts' group (SCCEG), has been set up to define the parameters of a half-rate speech and channel coding technique which will potentially double the system capacity.

From time to time ad hoc sub-groups and joint groups have been formed, and in accordance with the original terms of reference, liaison with other CEPT and CCITT bodies has focused their efforts on the needs of GSM. At one time there were 28 such groups and it was estimated that throughout Europe many hundreds of individuals were concurrently working directly on the GSM standard. These groups have been responsible both for the writing of the standard and the theoretical, practical and simulation studies underlying it.

In the case of the radio sub-system and the speech codec, the actual research, development and testing of specific practical techniques has been the direct responsibility of the WPs. This has been a highly collaborative exercise and the support of industry, particularly in the latter years, has been invaluable.

13.4 System features

We now turn to the second question - what are its features?

In detail the answer lies in 161 Documents, the GSM Recommendations [4], having a grand total of nearly 6000 pages, which result from the deliberations of GSM. In this chapter it is clearly not possible to do more than provide pointers to this enormous amount of information but the factors which both facilitate and restrict the features will be outlined. While a few of the Recommendations are descriptive reports, the majority are mandatory requirements to ensure that any mobile station will work in any GSM network and that infrastructure equipment purchased from different manufacturers will interwork correctly.

The recommendations are arranged under twelve main headings shown in Table 13.1.

01	General
02	Service aspects
03	Networks aspects
04	MS-BS interface and protocols
05	Physical layer on the radio path
06	Speech coding
07	Terminal adaptors for MSs
08	BS-MSC interface
09	Network interworking
10	Service interworking
11	Equipment specification and type approval
12	Operation and maintenance

Table 13.1   GSM recommendations

However the features which will ensure success in the market place are those which offer tangible benefits to users, and which are addressed only indirectly by the recommendations.

Those features will include the following, not necessarily in any order of priority.

- Good speech quality

- Low terminal cost

- Low service cost

- International roaming

- Ability to support hand-held portables

- A range of new services and facilities

Among the new services and facilities can be listed supplementary services and various non-speech services. In fact it was envisaged from the outset that the GSM system would have ISDN capability although the bandwidth of the mobile radio link restricts the maximum data range that can be used. Full ISDN numbering is, however, a feature of the GSM system which has been described as providing an ISDN terminal in the car. Such a description, however, tends to undervalue the importance of the hand-held cellphone which many operators now believe may eventually dominate the market.

Indeed this is underlined by the emergence of new mobile services relying on light pocket-sized terminals, such as telepoint and personal communications networks (PCN) which while offering public telecommunications service, nevertheless impose some restrictions such as one-way (outgoing) calling or speech-only service.

The GSM system, however, is a two-way system, is not restricted to telephony or speech-based services and offers data at rates up to and including 9.6 kbit/s, and two messaging services - Short Message Service Point to Point (SMSPP) and Short Message Service Cell Broadcast (SMSCB). Facsimile service is also offered, although it must be remarked that the timing requirements of Fax Group 3 (which is an analogue service while the GSM system is digital) the error performance of the mobile radio link and the potential need for retransmission impose some severe technical conditions.

Low service cost is related to the cost of providing the infrastructure and, like low terminal cost, is strongly influenced by the complexity required by the recommendations. While these are admittedly very detailed, nevertheless it is left to the ingenuity of industry to devise and develop low cost equipment. In order that such equipment can be supplied competitively by industry the functions and interfaces must be precisely defined. It is these which determine technically how the GSM system will work.

13.5 Interfaces

We now come finally to the third question - How does it work?

GSM has concentrated on defining the interfaces between the various components of the system so far as is necessary to ensure their correct functioning and also that satisfactory interworking with fixed networks can be provided. Only in this way can the GSM system function as a truly public telecommunications system.

When GSM started its work much of the effort was concentrated on radio systems. This reflects the essential, and indeed only real, difference between a mobile network and a fixed public switched telecommunication network (PSTN); that is, the substitution of the copper-pair subscriber's loop by a mobile radio path.

It is therefore appropriate that attention is focused on the radio aspects and these highlight the differences; however, in order to illustrate the similarities with fixed networks, the network aspects of the system are also considered.

Although there are several more interfaces, and they are all important for the operation of the system, only those relating to the radio and network access are discussed in any detail here. Other interfaces include, for example, the subscriber's interface module (SIM) which personalises a mobile station (MS); some MSs will have a slide-in smart-card SIM and some a wired-in one that is not user-changeable. Fig. 13.5 shows the more readily apparent interfaces.

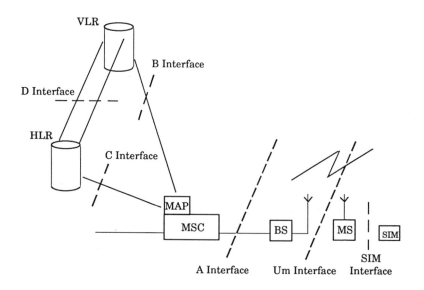

Fig. 13.5   Some interfaces of the GSM system

13.6 Network aspects

A PSTN has fixed geographic features that are taken for granted; only when mobility is considered do the limitations of these features become apparent. The routeing and switching functions of a PSTN exist to overcome the economic and practical difficulties of providing connections between every possible pair of terminals, but these functions and the copper pair to each telephone terminal then fix the relationship between a telephone number and its physical location.

However, because of the very mobility that characterises a PLMN this direct relationship is missing; a mobile station is mobile and does not have a fixed location, thus before it can receive calls its location must be determined. In fact the PLMN must have means for registering and authenticating, as well as locating, MSs. These means constitute data bases which manage the mobility of the MSs and their facilities and hence give a PLMN the characteristics of an intelligent network; the GSM system is probably the first such network to be defined completely.

When the MS is switched on three actions take place: it registers with the network, is authenticated and its location is up-dated. As it moves across cell boundaries, cell reselection takes place automatically; the network continuously tracks the MS at all times. Thus the network is able to route incoming (mobile terminating) calls quickly and efficiently, and permit or deny outgoing (mobile originating) calls according to the status of the MS's subscription, whilst calls in progress can continue without interruption as the MS crosses cell boundaries.

A GSM PLMN will have at least one home location register (HLR), where all management data relating to all home MSs is stored, and at least one visitor location register (VLR), where selected data relating to visiting MSs is stored. This data includes the international mobile station identity (IMSI), the MS international ISDN number, and other information including the current location of the MS. The HLR and VLR may be co-located with the mobile services switching centre (MSC) or not as the system operator decides and, of course, the number of HLRs and VLRs will depend on the size of the network. Thus the interfaces between the MSC, the HLR and the VLR must also be defined (see Fig.13.5).

Every GSM MS will be allocated to a specific HLR (the code of which forms part of the IMSI) and this information is used to enable calls to be made when, for example, a MS visits a network other than its home network. The MS is requested to pass the IMSI on the uplink to the nearest base station (BS) and this is forwarded via the MSC to the VLR of that network. The VLR accesses the HLR (via fixed network links, international if necessary) in order to obtain the selected information needed for registration.

A MS Roaming Number (MSRN) and a temporary mobile subscriber identity (TMSI) are allocated by the visited network. An authentication process is carried out; if this is successful access is permitted and MS originated calls can then be made while the registration of the MS's location enables incoming calls to be routed to the correct area.

Among the selected data transferred from the HLR to the VLR will be the services available to that MS, for example, supplementary voice services or data services. At the time of initiating a call specific details of the services requested will then be required from the MS.

From this very cursory description it can be seen that the signalling activity both on the radio path and on the fixed links of the PLMN is considerable. It can also be recognised that there is a large amount of data processing associated with the mobile management in addition to that required for normal call management and supervision.

The international signalling system CCITT No. 7 (see Chapter 8) has been specified for the network management and interconnection signalling functions of the GSM system and the requisite mobile application part (MAP - GSM Recommendation 09.02) has been defined by one of the CEPT groups in liaison with GSM. It is one of the largest of the GSM recommendations and runs to nearly 600 pages.

13.7 The radio sub-system

The radio sub-system (RSS) comprises the various aspects of the interface between base stations and mobile stations - the mobile radio link. Often called the air interface, it includes the definitions of the logical (i.e. the traffic and control channels) and physical channels (i.e. the radio frequencies and timeslots), the multiple access, multiplexing and timeslot structures, frequency hopping, coding and interleaving, modulation, power control and handover, synchronisation and transmission and reception.

Also included are the specifications of spurious emissions which could potentially have an adverse effect on other services. These arise from the modulation process and the power ramping at the beginning and end of each timeslot as well as the usual imperfections in the actual implementation of the oscillator, frequency synthesiser and other radio frequency elements that go to make up a mobile radio station.

All the physical (Layer 1) radio aspects are covered in the RSS recommendations (05-Series, Physical Layer on the Radio Path) and are accommodated within fewer than 200 pages, whereas the Layer 3 functions relating to the MS-BS interface are defined in the 04-Series (and require many hundreds of pages for their full specification).

13.7.1 Transmission and reception

Very early in the life of GSM it was difficult to get general agreement on the use of hand-held portable (HHP) MSs as some operators were reluctant to offer them; they are now seen as a significant factor in the market place which is important as they have a strong impact on system design. HHPs are now established as a firm requirement in the system and it is variously estimated that they will represent between 20% and 80% of all MSs. Agreement was an essential step because the disparity in RF power output between vehicle mounted and hand-held MSs has a significant impact on the design of the radio sub-system, in particular the path loss budget.

For example a Class 1 MS (typically vehicle mounted) has a peak power output of 20W or 43 dBm whereas a Class 4 hand-held portable has a peak power of 5W (37 dBm) and Class 5 only 2W (33 dBm). The maximum receiver sensitivity of both BSs and vehicle mounted MSs is -104 dBm while that of hand-held portables is -102 dBm. Hence it can be seen that when the path loss reaches a limiting value for the hand-held portable it is at a disadvantage compared with vehicle mounted MSs which still have several dB in hand.

This can be expected to occur most often in rural areas where the cells are typically large and there may be holes in the coverage. In these situations hand-held portable MSs may not give such good service as the more powerful MSs in classes 1-3 but they will perform excellently in the small cells that typify dense urban and city areas and where there will be much greater numbers of users. In such city areas with a large number of small cells the maximum BS power output is unlikely to be used and while this does not equalise the path loss it ensures that hand-held MSs are used well within their capabilities in most cases. The exception to this is the use of hand-held MSs inside buildings, lifts, underground railway stations and other situations where radio penetration from a conventional mast top BS antenna is relatively poor.

13.7.2 Radio path system organisation

The traffic (speech or data) and control signalling information transmitted digitally over the radio path are organised on a frame and multiframe basis in accordance with normal time division multiplexing techniques and conveyed on radio frequency carriers having a bandwidth of 200 kHz.

There are several types of control channel. The main ones are: a broadcast control channel (BCCH) on the downlink only to keep MSs aware of the base station identity, frequency allocation and frequency hopping sequence; a common control channel (CCCH) further subdivided to provide a random access channel (RACH) on the uplink, a paging channel (PCH) and an access grant channel (AGCH) on the downlink; a dedicated control channel (DCCH) for registration, location updating, authentication and call set-up; two associated control channels (ACCH) continuous stream (slow ACCH - SACCH) for call supervision and burst stealing mode (fast ACCH - FACCH) for power control and handover.

Information is conveyed in an 8-timeslot frame; traffic channels (TCH) are organised into a 26-frame multiframe of which 24 frames carry traffic while the remaining two carry the SACCH; when needed, the FACCH steals a TCH. Recommendation 05.02 gives full details of all the structures. The structure for the traffic channels is shown in Fig. 13.6.

The BCCH is organised into a 51-frame multiframe and is carried in timeslot zero on a non-hopping radio frequency carrier; the remaining timeslots are available for traffic channels. Several radio frequency carriers are allocated to a BS of which one will carry the BCCH. The use of hopping or non-hopping radio frequency carriers for the traffic channels is optional.

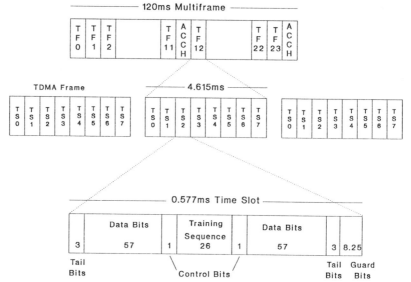

Fig. 13.6   Frame and multiframe structure on the radio path

13.8 Speech coding

The GSM system is a digital system and in order to exploit digital techniques effectively the speech signals must be encoded at a low bit rate; the digital techniques used in the PSTN, e.g. pulse code modulation (PCM) at 64 kbit/s or adaptive differential pcm (ADPCM) at 32 kbit/s, would not use the radio spectrum efficiently enough within the restricted allocations to provide the number of channels necessary to support a high capacity system.

The speech coding rate deemed possible was related to the then current state of speech coding technology and was fairly conservative. GSM recognised that when the system opened for service it was highly likely that lower-rate codecs (codec = coder-decoder) would be available and therefore made provision for them to be incorporated into the mature system later.

Thus the nominal rates of 16kbit/s for the full-rate early system and 8 kbit/s for the later, half-rate system, were agreed. The maximum delay introduced by the codec was also defined together with the TDM framing delay. The reason for this is that experience has shown that delay has a direct effect on customers' satisfaction with the circuit; there is ample evidence of this from satellite communications.

Unfortunately this turned into a political debate since one candidate speech codec had very low delay (about 20 ms) while most others had much larger delay (>50 ms). Some GSM delegates felt that satellite delays of 260 ms demonstrated that delay was not a problem and it was difficult to convince them otherwise. However, the tandem connection of a GSM mobile call with a satellite connection would exacerbate the problem and there is still no practical way of preventing this from occurring. Eventually, upper limits of 65 ms for the codec and 65 ms for TDM framing making a total of 130 ms were set; development has produced an overall delay of about 90 ms (making the round trip delay 180 ms) of which the speech encoder accounts for 25-28 ms.

The net bit rate which finally emerged was 13.0 kbit/s. The method chosen employs residual pulse excitation (RPE) linear predictive coding (LPC) with the addition of a long term predictor (LTP). However, this codec has relatively little inherent redundancy and due to the impairments of mobile radio propagation including slow and fast fading together with noise and interference it is necessary to protect the encoded speech by means of channel coding.

13.9 Channel coding

Channel coding techniques were investigated to determine the right balance to ensure an adequate level of protection and adequate data rates for the various services. The technique finally chosen for speech employs a convolutional code of rate 1/2 block-diagonally interleaved over 8 TDMA frames to provide protection against burst errors.

The 20ms speech frame consists of 260 bits; a 3 bit cyclic redundancy check (CRC) is added to the 182 most significant bits making 185 bits, 4 tail bits are then added (to initialise the decoder) to produce 189 bits which are then presented to the 1/2 rate channel encoder producing 378 bits. The remaining 78 least significant coded speech bits, which are unprotected, are then added making a total of 456 bits.

The bit rate of the resultant protected channel is therefore somewhat higher than that of the speech encoder output at 22.8 kbit/s (456 bits in 20 ms) but the traffic channel is now fairly robust and is capable of providing good communication under adverse conditions.

The effect of errors on the signalling, system control messages and data are not the same as for speech so different coding rates are employed in order to provide a balanced degree of protection.

13.10 Multiplexing and modulation

The TDMA frame is produced by multiplexing together eight channel-encoded speech sources in time division as shown in Fig. 13.6. Eight timeslots of duration 0.577 ms make up one TDMA frame of 4.615 ms and the transmitted bit rate on the radio path is 270.83 kbit/s (156.25 bits in 0.577 ms approximately).

However, simple arithmetic applied to the 8 channels each of 22.8 kbit/s gives a false result since it does not take account of the training sequence, the guard periods and the slow associated signalling channel which account for the remainder. This demonstrates graphically the impact of the system requirements on the transmitted bit rate amounting to nearly one-third of the total.

By the use of Gaussian minimum shift keying (GMSK) modulation with a BT (3dB bandwidth x bit period) product of 0.3 the spectrum of this signal is tailored to enable it to be conveyed on a radio frequency carrier of 200 kHz bandwidth.

As the total available radio spectrum is planned to be 2 x 25 MHz it can be seen that this radio frequency channel allocation allows for 125 radio carriers. Each carrier conveys 8 time-divided channels making a total of 125 x 8 = 1000 channels.

Since the current UK analogue TACS system uses 25 kHz channels (allowing 1000 channels in 25 MHz) it is relevant to ask how the GSM system is to achieve high, or more specifically, higher, capacity than TACS.

Because of the inherently greater robustness to interference of the protected digital signal the GSM system is designed to operate at a much lower channel to interference (C/I) ratio of 10-12 dB compared with 17-18 dB for TACS. Tests have confirmed that a C/I ratio of 9 dB is attainable (including an implementation margin of 2 dB) giving the GSM system an advantage over TACS of 8-9 dB. By this means a smaller re-use distance can be employed leading to smaller cells thus giving the system higher capacity.

One of the objects of the validation trials has been to establish the operating C/I compared with the original target figure and it can now be seen that the system is expected to have a better performance than the target. Of course, it still remains to be seen what effect this C/I has on the bit error ratio (BER) and hence on the speech quality in a fully loaded system.

13.11 Multipath propagation and equalisation

Multipath propagation is a characteristic of mobile radio and one of its effects is that delayed signals combine to give a composite signal at the receiver antenna whose amplitude and phase undergo wide and rapid fluctuations. This subject is treated extensively in Chapter 2 so only a simple reference is made here. The effect can be understood by reference to a simplified two-path model, Fig. 13.7; depending on the length of the second path the delayed signal may arrive late enough to cause serious intersymbol interference thus limiting the maximum usable bit rate.

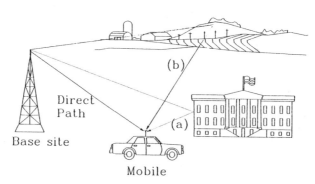

(a) Short Reflected Path
(b) Long Reflected Path

Fig. 13.7   Illustration of multipath propagation

In practice there are many such paths and as the MS moves it will be subjected to a rapidly changing multipath propagation environment. In the GSM system the long excess delays resulting from signal components whose paths are very long can only be tolerated by using an equaliser in the receiver.

A known bit pattern, or training sequence (26 bits), is transmitted at regular intervals and the equaliser compares the received bit pattern with the training sequence adjusting the parameters of a digital filter so as to produce the inverse transfer function to that of the radio path. As the changes are rapid and because the timeslots are independent every timeslot must contain the training sequence.

This is an extra overhead on the information carrying capacity of the digital channel and further reduces the effective throughput. Without equalisation, however, the effective range would be limited to perhaps a few hundred metres where these effects are less serious; on the other hand a cheaper mobile station with no equaliser is conceptually possible for very short range operation, for example in microcells.

Mobile stations have to operate in widely differing propagation environments, for example cities with high rise buildings and virtually no line-of-sight paths, rural areas and open waterways with frequently clear line-of-sight paths (but also possessing many reflected paths), and mountainous areas with long excess delays. They must also operate at widely varying transit speeds in cars on city roads and motorways, in stationary cars in laybys or in traffic jams, and when carried by pedestrians in all kinds of situations, including on board high speed trains.

Clearly the characterisation of these conditions is complex and the design of the equaliser performance is complicated by such a variety. In the GSM system the design of the equaliser is not specified and is left to the ingenuity of the manufacturer but its performance is defined in terms of a maximum excess delay of 16 µs. The definition is quite complex and is illustrated by the type approval conditions.

Type approval is subject to conformance with a set of tests using a system simulator defined in the 11-Series recommendations. This makes use of a propagation simulator which models a number of multipath profiles together with appropriate Doppler spectra. Typical urban, rural and hilly terrain conditions are specified as also are typical vehicle speeds. A vehicle is anything which physically transports the MS, including a pedestrian, so causing it to move in relation to a base station thereby creating a non-stationary signal reception pattern. The equaliser must perform satisfactorily so that the resultant error performance is within specified tolerances.

13.12   Radio link management

The radio link, comprising both uplink and downlink, has to be managed in order to ensure continuity of service and to minimise interference to other users of the system. This is a complex topic but the main points will be addressed.

13.12.1   Timing advance

The propagation time of the radio signals between the MS and the BS will be determined by the distance that the radio signal travels. Since cells in the GSM system may vary in size from perhaps 1-2 km up to about 35 km it can be seen that the propagation time can vary from about 3-100 microseconds. In order that the data bursts transmitted by each MS fall exactly into the timeslot structure at the BS receiver it is necessary to advance the timing of the MS transmitter by an appropriate amount and this must be done individually for each MS. This is done at call set up by the BS which measures the round trip time and sends a timing advance message to the MS which is amended as necessary at intervals during operation.

13.12.2   Adaptive power control

Adaptive power control is applied to all MSs to ensure that they operate at the lowest power level consistent with adequate receive signal strength and quality. The power is controlled in steps of 2 dB from the maximum defined by the power class.

For example a Class 1 MS (peak power 20W or 43 dBm) has 15 steps giving a minimum power level of 13dBm. The other classes are catered for by the same algorithm since their peak powers, 8W (39 dBm), 5W (37 dBm) and 2W (33 dBm) all correspond to steps on the same scale.

Power control is achieved by a process which involves measurement by the MS of the received signal strength and quality and the regular reporting, on the uplink, of these data to the BS. The BS has preset parameters to enable a decision to be made and when a threshold is reached it commands the MS to change power level either upwards or downwards as necessary. The received quality is a measure of the BER.

Base stations will have an adjustable peak power level to allow the system operator to make adjustments to the cell coverage area and in addition they may employ power control on the downlink in a similar way to that defined for MSs.

13.12.3   Handover

This is one of the basic features of cellular radio and is provided to enable MSs to move freely across cell boundaries and have continuous service; i.e. by means of it calls in progress are maintained across cell boundaries. It uses the same process of measurement and reporting as for power control; indeed the two procedures are closely linked.

The GSM system exploits the properties of TDMA very effectively for handover; the MS listens for the BCCH of up to 16 surrounding BSs in the timeslots that are not being used for transmission or reception. It forms a list of up to 6 handover candidates and reports the signal strength and quality to the serving BS. Meanwhile the BS is also monitoring the signal strength and quality on the serving channel. Handover is under the control of the network and is used to provide continuity of communication because of the availability of another channel which can allow communication at a lower power level. Handover can also be used for traffic balancing between cells.

Handover to another cell will involve retuning to another radio frequency channel, but handover is also possible to a different timeslot on the same radio frequency channel in the same cell; this may be used for interference control reasons.

13.13   Measures to optimise performance

As with all cellular systems the GSM system will operate under interference limited conditions. With the expected high capacity of the system correspondingly high levels of co-channel interference can be predicted. Various additional measures have been introduced to control this and to allow the system to grow to the fullest extent possible and these can be seen as refinements to the radio subsystem.

13.13.1   Frequency hopping

As multipath propagation is frequency dependent and since several radio frequency channels are available in any cell the technique of slow frequency hopping can be employed to mitigate its worst effects and this is particularly effective for slowly moving and stationary MSs.

The improvement in performance given in this way has been variously estimated at as little as 1-2 dB and as much as 6 dB and this is another aspect where the validation trials will shed light on the actual performance to be expected.

The complexity introduced by frequency hopping is not great but in the GSM recommendations it is mandatory only for MSs and network operators may opt to use it or not at BSs.

13.13.2   Discontinuous transmission

To minimise still further the levels of interference caused by MSs discontinuous transmission (DTX) will be employed. This relies on voice activity detection (VAD) in the speech encoder and switches off the MS transmitter when the speech encoder has no output data for transmission. This has an additional benefit since it minimises the power drain on the battery in hand-held portables. DTX can also be used at BS at the option of the system operator.

13.13.3   Discontinuous reception

A further mechanism for minimising this power drain is the use of discontinuous reception (DRX) sometimes known as sleep mode operation. By this means the MS receiver periodically switches out of idle mode for very short intervals in order to receive messages on the downlink. These may be, for example, paging messages relating to an incoming call, or system control messages. It is important that the duty cycle of DRX allows the MS to respond to paging messages in an acceptable time for handling incoming calls (and also SMS messages). The duty ratio depends on the number of paging channels and is of the order of 2-9% and this greatly eases the load on the battery.

13.14   Conclusion

The GSM recommendations provide a framework but do not provide answers to all of the issues, in particular the important steps listed below. These are not addressed and require a great deal of engineering judgement based on a detailed understanding of the Recommendations and it is here that competitive advantage is to be won or lost.

- Procurement specification

- Tender

- Contract placement

- Delivery

- Trial system

- Operational network

This chapter has attempted to answer three questions about the GSM system;

- What is it?

- What are its features?

- How does it work?

Of necessity much of interest has had to be omitted, for example the work of COST in evaluating digital modulation techniques and radio propagation factors, and the details of the evaluation of the various candidate speech codecs and radio system proposals. However, the system is extremely complicated and it would take a whole book to cover all the aspects in depth; even a guide to the 6000 pages of the Recommendations would be longer than this chapter.

At the time of writing the recommendations are in the process of being subjected to the Public Enquiry procedures of ETSI and Interim European Telecommunication Standards and many readers will no doubt be involved in that process. For those who would like their own copies the Recommendations are available from ETSI PT12, on payment of a small charge, at the address given in Reference 4.

This chapter has attempted to provide an appreciation of the system to enable designers and users to derive the maximum benefit from this European venture. The volume of literature is a telling testimony to the truly international effort involved in defining and specifying the first international public mobile telecommunications system, the GSM system.

The pressure on radio spectrum, the increasing involvement of industry and the initiative to increase capacity by means of half-rate speech coding even before service opens all serve to emphasize the fact that mobile communications is one of the fastest growth areas in telecommunications today.

As remarked earlier, in 1982 the outcome of the work just starting on the European-wide PLMN standard was visualised in only a general way. Perhaps, if the sheer size of the task had been realised fully in those early days, the will to achieve it would have been lacking. In which case we would not have arrived today at the threshold of the most comprehensive telecommunications system ever to have been devised as a complete entity. The achievement of that outcome in 1991 may be not only the result of foresight - but also, ironically, fortuitous lack of vision.

13.15 References

1. Aide-mémoire: Conference Européene des Administrations des Postes et des Télécommunications (CEPT), Berne, November 1982

2. Advanced Mobile Phone Service: Bell System Technical Journal, Vol. 58, No. 1, January 1979

3. United Kingdom total access communications system; mobile station - land station compatibility specification, Issue 4, Aug. 1989: DT I, Victoria St, London

4. ETSI/GSM Recommendations: ETSI, BP 152, F-06561 Valbonne CEDEX, France

Chapter 14

The pan-European cellular technology

David M Balston

14.1 Introduction

Radio systems are becoming more and more complex as designers strive to provide a quality of communications and a reliability of service which matches that offered by the fixed telecommunications networks. Furthermore in contrast to the fixed networks, it is just not possible to increase capacity by laying more cables and the problem is made more difficult by the system bottleneck caused by the limited spectrum availability. The pan-European cellular system, also known as GSM (Groupe Spéciale Mobile), is considered by many to be the most complex radio system in the world, including military networks. Those of us working to implement the system consider that we have to achieve military complexity at consumer prices, in a timescale which would make a soldier green with envy. The job is made even harder by the fact that the specifications are still not complete and we are targetting to get our customers, the network operators, in a position to open full service by July 1991. In this chapter a description of some of the technical problems which face implementing the system are set out with the purpose of demonstrating that the manufacturers' confidence in the system is not misplaced.

There are four categories of technology which impact greatly upon the GSM development programme,

- RF
- Digital signal processing
- Software
- Control algorithms

It is most important to emphasise the last two because, whereas the word technology tends to be associated with VLSI and clever circuit techniques, one must also consider the wider implications of controlling the network.

Before discussing each of these topics in detail, there are a few general points which tend to cut across all topics and which need to be taken into account by the engineering teams.

14.1.1 Timescale

By any standard the development schedule being worked to is tight. We have three responses to this. Firstly, in common with most manufacturers, we participate fully in the GSM specification process. This gives advanced warning of the issues, the likely outcome and the ability to influence the decisions. Secondly to have equipment working early, we must prove our feasibility and feed back to the specification teams the results of our endeavours. Thirdly we have to validate the specifications, prove that the system performance can meet targets and transfer this experience into product design teams.

14.1.2 Power consumption

The UK cellular phone market has seen a very significant emphasis on the hand portable unit. Approximately 25% of all cell phones sold today in the UK are portable and we are expecting this trend to continue. This places a strong demand on the need to minimise the equipment power consumption in order to keep the battery size down commensurate with reasonable standby and talk time figures. This in turn implies greater silicon integration, power saving device architectures and eventually low voltage supply rails.

14.1.3 Technology intercept

GSM has been designed to capitalise on the sub-micron VLSI technology which will emerge over the next couple of years. Although very complex, the digital signal processing needed will already fit on only three chips and the prices of these should fall dramatically as the level of integration goes up, the die sizes fall and the market volume takes off.

14.2 RF Issues

14.2.1 Digital modulation

14.2.1.1 *GMSK modulation*
The GSM specifications require the out-of-band radiated power in the adjacent channels to be between 40 and 70 dBs below that of the desired channel. To satisfy this requirement, it is necessary to bandlimit the RF output signal spectrum. Such a spectrum manipulation cannot easily be performed at the final RF stage in the multichannel transceivers. In order to mitigate the filtering problem narrow band digital modulation schemes must be used.

Amongst these schemes, Minimum Shift Keying, (see Chapter 4) has proved to provide a better modulation for use on bandlimited and nonlinear channels. It has the following properties:
- (a)   constant envelope
- (b)   relative narrow bandwidth
- (c)   coherent/non-coherent detection capabilities
- (d)   a good bit-error-rate performance measure

However MSK does not completely satisfy the out-of-band radiation requirements and to overcome this limitation a premodulation low-pass filter is used. A Gaussian filter is suitable for this role and the modified modulation is termed Gaussian MSK (GMSK).

GMSK is a form of FM. The carrier is therefore at a constant amplitude during transmission, allowing the use of a low cost, efficient class C transmitter.

GSM has set very tight specifications for the accuracy of the GMSK modulator, both in terms of frequency domain sidebands and time domain phase trajectory. The reason for this is to ensure compatibility between mobiles and base stations throughout Europe and ensure correct operation of the various error correction procedures envisaged. These specifications coupled with the data transmission rate of 270 kb/s present a serious challenge to the equipment designer.

There are essentially four different ways of modulating a carrier for GSM.

14.2.1.2 *Direct digital synthesis (DDS)*
This is the most accurate method available. DSP techniques are used to address a ROM containing a sinewave lookup table. The output from the ROM is fed to a fast DAC to directly generate a modulated carrier. DDS technology is developing rapidly, and carriers at over 100MHz can be generated. The solution is however at present very expensive and a further frequency shift to 900 MHz would still be necessary.

14.2.1.3 *IQ modulation*
This is the 'traditional' method. The data stream is converted to a baseband Cartesian representation and then mixed up to final frequency using an IQ modulator.

The technique is prone to problems with residual AM due to imbalance in the mixer system. If a class C transmitter is used the AM gets converted to phase modulation and the accuracy of the modulation is compromised.

14.2.1.4 *FM modulation*
Phase locked loop synthesizers are the most commonly used method for generating the carrier in mobile radio systems. FM can be introduced into such a loop at various points, and each has its own problems in terms of accuracy, stability and bandwidth.

The ideal technique is to use a high bandwidth, high resolution synthesizer to achieve the modulation directly by selecting different frequencies at some multiple of the data rate. In the case of the base station however the data rate can have a component at or near dc and the loop bandwidth would not be able to cope with this.

14.2.1.5 *Direct digitalinterpolation*
A solution which overcomes most of these problems is offered by a technique which can best be considered as a combination of direct digital synthesis and phase locked loop synthesis. The details of this approach are, however, commercially sensitive and no more details can be given at this stage.

14.2.2   Power control

In order to keep to a minimum the interference caused by transmitters in a mobile network, a particular traffic channel is generally operated at the lowest possible power level commensurate with the quality of communication required. This is achieved in GSM by the mobile modifying its output power wherever necessary under the control of the base station and the latter likewise adjusting its power level. GSM defines 16 power levels in the range 20 mW to 20 W and any one step must be accurate to ±3 db (and be part of a monotonic sequence). The maximum power level of any class of mobile must have an absolute accuracy of ±1.5 dB.

In both the mobile and the base station the TDMA system requires that the transmitter must produce pulses of power and these must be carefully shaped to avoid the generation of sidebands which may cause interference to adjacent channels. Whereas the mobile needs to ramp its power up from zero each time, the base station can be transmitting 8 successive timeslots, each on a different power level. Thus the ramping control has to accommodate a very wide range of level changes.

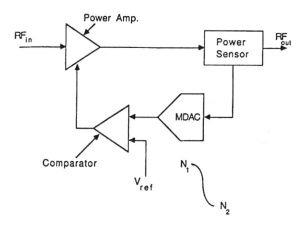

Fig. 14.1   Power control solution

Fig. 14.1 shows one solution to the problem. The voltage-controlled power amplifier is fed by a signal which has been derived from the output and modified by a control signal applied to a multiplying digital-to-analogue converter. This control signal is the inverse of the power ramp required (shown as the curve N_1 to N_2 in the figure) and as the comparator detects a change in the signal level at its input compared with a fixed reference voltage its output will change until stability is again attained. In this way the output power level will follow the control signal profile.

This approach automatically compensates for the considerable nonlinearity of the gain curve in the power amplifier and this compensation is extended to any changes in gain caused by temperature or aging effects. Care must be taken to achieve a very stable reference voltage however.

Other problems associated with this approach are temperature stability of the sensor and glitches on the DAC caused by discrete changes in the control signal which can become amplified in the control loop and give rise to substantial noise.

14.2.3 Frequency hopping

GSM mobiles have to monitor the signals from those base stations surrounding the one they are using in order to provide information to the network on the need/ opportunity for hand-off. By definition these adjacent stations will all be operating on different frequencies. The mobile monitors adjacent base stations during the dead period between transmit and receive in one time frame. Thus within the period of 4.6 ms the mobile has to receive a pulse, transmit a pulse and monitor another signal - all on different frequencies. This inherent frequency agility can however provide an extra benefit if the base station itself can operate in a frequency agile mode.

Frequency hopping is a technique normally applied to military radios, and refers to a means of evading deliberate jamming or interception of a mobile radio link. In a normal commercial application however the technique can be used to avoid loss of communication due to bad RF channels, whether caused by non-deliberate jamming or bad multipath effects.

Without frequency hopping a mobile parked in or travelling through a fade would not be able to achieve or maintain communication. Fading is a narrow band phenomenon however and thus switching to a different frequency would allow communication to be maintained. In practice there is sufficient redundancy and error protection in the GSM signal structures to allow one frame in 5 to be lost without significant loss of quality.

Although frequency hopping is common in military radios, it has never been implemented in commercial equipment. The challenge to the designers is to create a frequency synthesis solution which can achieve 2000 hops per second, can settle in 100 µs (30 µs for the base station) without compromising the modulation phase trajectory, and is also small and cheap to implement.

Traditional solutions have adopted two or more synthesizers to meet the settling time requirements, one synthesizer being used while the other is settling to the next frequency. This can be an expensive solution and clearly takes up more room. It is also difficult to totally suppress the output of the non-active synthesizer as it sweeps across the frequency band. The direct digital interpolation synthesizer is however capable of meeting these requirements for the mobile.

For the base station effective frequency hopping for all of the traffic requires a minimum of five carrier frequencies to be used. The carrier containing the control channel information must be fixed in GSM and thus if less than five carriers are available, hopping can only occur over three frequencies. This is too small a number to give an adequate benefit.

These conditions led us to choose baseband hopping at the start of base station design work. In this system the transceivers do not change frequency, the baseband signals are switched between fixed frequency equipments.

14.3 Signal processing

14.3.1 Digital speech coding

The codec selected for GSM is a Regular Pulse Excitation, Linear Predictive Coder (RPE-LPC). This algorithm is based upon a general fixed point 16 bit DSP core. The algorithm has now been successfully implemented on many different DSP chips e.g. Analogue Devices ADSP2100, AT&T DSP16, Motorola DSP56000 and Texas Instruments 320C25.

The basic data rate from the codec is 13 kb/s and speech is processed in 20ms blocks.

The speech codec also has to support several other functions:-

- Voice activity detection to allow the use of *Discontinuous Transmission*. This can save battery power in the mobile because the mobile will not transmit when the user is not speaking. DTX also reduces co-channel interference which allows greater frequency re-use and hence more users for a given resource.

- Comfort noise generation to prevent an annoying 'dead' sound to the receiving party during discontinuous transmission. The receiving codec will generate comfort noise that tracks the background noise of the calling party.

- Speech extrapolation to mask the effect of speech frames received with errors from the radio. Here the decoder must repeat the decoding of the previous good frame with a little attenuation.

One significant advantage inherent in the digital transmission of speech is increased security from eavesdropping and if more security is required then digital encryption may easily be applied.

The design of a GSM codec would be fairly straightforward if size, power consumption and cost issues were not important.

The design of a cost effective solution for use within the mobile environment however gives rise to two major areas of difficulty. Firstly, we need a 13 bit linear analogue-to-digital converter with appropriate analogue antialiasing filters, together with a complementary 13 bit digital-to-analogue converter. Secondly, the reduction in size and power requirements of the DSP core and the incorporation of significant amounts of digital interface circuitry and high speed RAM and ROM. Both these problems require the intimate support of the DSP manufacturer and in common with our competitors we are working closely with such a silicon manufacturer to establish a suitable solution.

14.3.2 Multipath equalisation

The data rate selected by GSM is 270 kb/s. The TDMA approach adopted was chosen to reduce, as far as possible, the amount of radio equipment required for each traffic channel. The wider the bandwidth and the higher the data rate the more users can share the same radio frequency and associated equipment. There is a limit however to the data rate that can be realistically employed and this is governed by the effects of multipath. At the transmission frequencies used for GSM, signals are easily reflected from buildings, high-sided vehicles and even hills and mountains. By definition a reflected signal will arrive at the receiver later than the direct path signal, because it has to travel further and GSM has been specified that receivers have to successfully demodulate signals which have been delayed by up to 4 bit periods. The process undertaken to achieve this is termed multipath equalisation.

In order to understand how the equalisation process works we need to discuss the effect that multipath has on the the GMSK modulation technique employed by GSM. This form of modulation is best considered as a constrained phase modulation. The carrier phase is changed by $\pm\pi/2$ depending on the value of the symbol being transmitted.

In a perfect system with no noise or phase errors, the phase of the received signal would this be one of the four values 0, $\pi/2$, π, $3\pi/2$ and the amplitude would be constant. Both phase and amplitude of the received signal are however severely corrupted by multipath.

GSM helps the receiver overcome the problem by transmitting a fixed sequence of 26 bits known as the training sequence in the middle of every time slot. The receiver, knowing the sequence, uses a correlation process to find this bit pattern and the correlation function itself provides an estimate of the transfer function in the time domain between the transmitter and the receiver. Once the transfer function is known, a prediction of the received phase and amplitude can be made for each combination of bits that might have occurred (the number of bits contributing to the combination is dependent upon the multipath time delay range which must be accommodated). It is now possible to determine the most likely bit sequence corresponding to the received signal by taking all possible bit sequences and defining for each the predicted phasor sequence. This can be compared with the received phasor sequence and a figure for the goodness of fit can be calculated. If we refer to Fig. 14.2 we see a received signal S_n. For simplicity this is assumed to have arisen from only two possible phasors, P_n and Q_n (in practice it could have come from any of the 16 possible phasors in the diagram). For the sake of argument let us assume that P_n represents the phasor which would have arisen from the sequence 10110 and Q_n that from 00100. The next bit to arrive will either extend the sequence to 101101 or 101100. Thus the only phasors which should be received next are those arising from 01101 and 01100. These are shown as $P_{n+1,1}$ and $P_{n+1,0}$ in the figure. A similar argument applies to Q_n creating $Q_{n+1,1}$ and $Q_{n+1,0}$. Thus if the signal S_n had arisen from P_n its successor S_{n+1} could only come from $P_{n+1,1}$ and $P_{n+1,0}$. We can measure how closely S_{n+1} fits $P_{n+1,0}$, $P_{n+1,1}$, $Q_{n+1,0}$ and $Q_{n+1,1}$ simply by measuring the distance it is from each. In practice we measure the goodness of fit between a signal and all predicted phasors for every bit interval. This allows us to construct a diagram known as the Viterbi trellis. This is shown in Fig. 14.3 and enables us to track all the possible routes through the trellis summing the costs C_{ij}^n at each stage. The sequence that has the lowest cost is then selected as the demodulated signal.

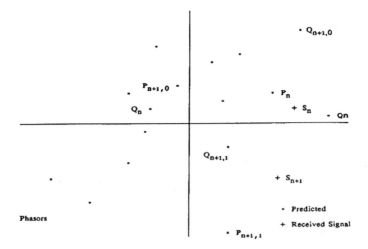

Fig. 14.2 Phasor prediction diagram

P_n	Predicted phasor from sequence 1 0 1 1 0
$P_{n+1,1}$	Predicted phasor from sequence x 0 1 1 0 1
$P_{n+1,0}$	Predicted phasor from sequence x 0 1 1 0 0
Q_n	Predicted phasor from sequence 0 0 1 0 0
$Q_{n+1,1}$	Predicted phasor from sequence x 0 1 0 1
$Q_{n+1,0}$	Predicted phasor from sequence x 0 1 0 0
S_n	Received signal at time n
S_{n+1}	Received signal at time n+1

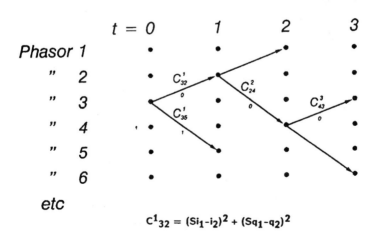

$$C^1_{32} = (Si_1 - i_2)^2 + (Sq_1 - q_2)^2$$

Fig. 14.3 The Viterbi trellis

The Viterbi equalisation algorithm has been successfully implemented in DSP devices, but Fig. 14.4 shows the processing times associated with them. Note that none of them in their present form meets the GSM requirements. They all also consume a lot of power and for the terminal a custom device is the only practicable solution.

TMS 320C25	17mS
ADSP 2101	9mS
DSP 32	5mS
GSM requirement	< 4.6mS

Fig. 14.4 Processing times for equaliser

14.3.3   Channel coding and decoding

The radio path will suffer from both random errors and fades, and GSM has adopted a convolution coding scheme to protect against these. Convolution coding has been shown to be superior to binary block code for random errors and to give similar performance to a Reed Solomon code for burst errors, provided suitable block interleaving is employed. Fig. 14.5 shows how the convolution coding with a constraint length of 5 is implemented. The particular encoder shown is that used for speech traffic. Different codes are used for signalling channels and data traffic.

Shift Registers

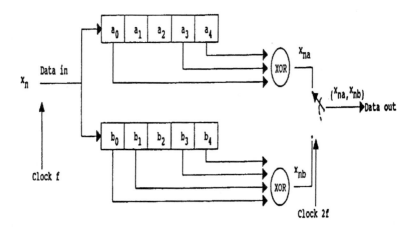

Fig. 14.5 A half- length convolution coder of constraint length 5

Decoding the received signal however is a much more complex process and the general consensus is that a Viterbi algorithm, very similar to that used for multipath equalisation, provides the best solution. Fig. 14.6 shows how this works. The coder outputs for possible combinations of data in the shift registers is established and the received bit pair compared with them. A simple Hamming distance cost function is then computed for each comparison - these are the figures shown below each received bit pair in the diagram. As the sequence proceeds the cumulative cost figure is determined and at the end of the block (182 bit pairs for a full rate speech channel) the sequence with the minimum cost function is chosen. For the sake of clarity only a small number of sequences are shown in this diagram but in practice all possible sequences will be evaluated.

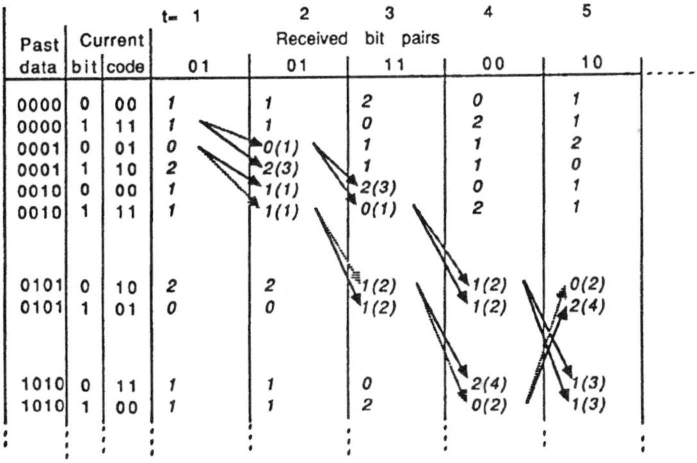

Past data	Current bit	code	t= 1 01	2 01	3 11	4 00	5 10
0000	0	00	1	1	2	0	1
0000	1	11	1	1	0	2	1
0001	0	01	0	0(1)	1	1	2
0001	1	10	2	2(3)	1	1	0
0010	0	00	1	1(1)	2(3)	0	1
0010	1	11	1	1(1)	0(1)	2	1
0101	0	10	2	2	1(2)	1(2)	0(2)
0101	1	01	0	0	1(2)	1(2)	2(4)
1010	0	11	1	1	0	2(4)	1(3)
1010	1	00	1	1	2	0(2)	1(3)

Fig. 14.6 The Viterbi decoder
(Figures in brackets are cumulative cost)

Note that it is not sufficient to select the lowest cost path at each point and only follow that route. In the diagram, for example, at time t=3 the decoded sequence 101 (shown by the dotted arrows) is winning but by t=5 the sequence 01010 (shown by the striped arrows) shows a lower cost figure.

It is possible to consider a refinement of this technique which makes use of additional information from the equaliser. Each bit in the sequence output by the equaliser can have associated with it a figure of merit based on the confidence with which the decision was made. If there was little difference between the zero or one paths at that point in the trellis then clearly the confidence in the decision would have been low. This figure of merit can be incorporated in the cost function at each step in the decoding trellis. It is predicted that this will give an improvement performance of about 2dB in a GSM mobile and make the C/I specification of 9dB achievable.

14.4   Software

The GSM system contains most of the features and control protocols of an ISDN system plus all the extras associated with a mobile radio network - radio control, tracking of mobile station location, handover, link security, etc. The mobile station alone is anticipated to require about 256 kbytes of software code. The network infrastructure in addition has to create and maintain large databases containing subscriber data and location, cope with all the fault detection and subsequent defensive actions needed to maintain service, modify its configuration to improve service, upgrade the software on occasion at every base station, etc. The base station subsystem part of the network (i.e. that part responsible for ensuring effective use of the spectrum in a limited geographic area) is estimated to require about 200 man years of software development.

14.4.1   Radio systems software

Clearly a system as complex as this requires a very disciplined approach to software development. The situation is not eased by the real-time nature of the problem and the tight execution time constraints of some of the functions. For example a mobile must:

Control hopping each TDMA frame	- say every 5 msec
Report radio power measurements	- say every 480 msec
Check identity of surrounding cells	- say every 30 sec
Action control messages from network	
Keep track of user input	

There are several ways in which one can elect to attack the problem in both the mobile and the base transceiver station by identifying a suitable existing real-time executive which gives full multitasking support and runs on a 16 bit microprocessor. In our case we have chosen VRTX for the mobile and Realtime Craft for the BTS, both running on 68000 processors. We have furthermore chosen to write code in Ada for the BTS as this demands more up front rigour than other languages and should reduce the integration problems downstream. We judged Ada to be a little heavy on overheads for the mobile (where again handportable power consumption considerations limit the amount of processing power available) and we are using 'C' in this case.

   Both software teams are following a rigorous SASD approach (System Analysis Systems Design) which first involves producing a model of the requirements showing the functions which have to be fulfilled and the relationship between them. Next a system design which meets this requirements model is derived and the traceability of a design element to a requirement is maintained throughout the design cycle. At this level it is possible to check the internal consistency of the design to ensure, for example, that the communication between two modules is specified in an identical manner in each. The next step, for those systems which will be implemented on more than one processor, is to partition the design onto the processors. Only at this stage are the specifications of individual modules established and coding and testing got underway. We are using a tool called Teamwork to support the SASD process. This provides all the necessary drawing and text editing facilities together with consistency checking and database routines to enable examination of the model from different perspectives.

14.4.2 Network management software

The GSM system has been specified to conform to the Intelligent Network concept IN/1 . Fig. 14.7 shows the equivalence of the node names in GSM and IN/1. The Network Management function (TMN in the diagram) must interact in real-time with all other nodes using standard OSI protocols. It is anticipated that knowledge based system techniques will be employed and thus the software structure must also support these.

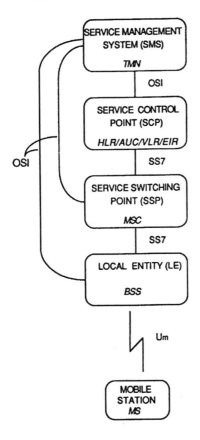

Fig. 14.7 The intelligent network concept - IN/1

There are therefore a number of problems facing the system designer:

- the software should preferably be portable to a variety of OEM processor platforms.

- it must cope with a variety of PLMN sizes.

- it should be flexible to allow evolution, both in the PLMN and the TMN.

- it must have a high integrity, i.e the failure of any part should not seriously damage the functionality or performance.

- it should preferably be a natural evolution from existing TMNs and it should be capable of interworking with them.

14.4.3   Data base software

The databases form the SCP (Service Control Point) of the intelligent network and have the added requirement that they have to provide an SS7 communications path to the MSC. The main task facing the system designers here however is how to provide a modular solution which will not be overly expensive when the network is small and yet can grow to provide the very fast transaction processing support needed on a large network. It has been estimated for example that a network of only 60,000 subscribers will require an HLR capable of handling 50 transactions per second (both single and composite).

14.5   Performance

14.5.1   Background

The two main factors which characterise a cellular mobile radio communications system, as distinct from any other, are frequency re-use across cells based on wanted carrier to co-channel interference (C/I) ratio tolerance and handover of calls between cells as users move across cell boundaries.

The actual C/I levels in the operational system strongly influence the system performance as seen by the user and are paramount in determining the spectrum efficiency (and hence the economics of operating a system). The GSM system is designed to operate with a C/I level of 9-10 dB which will support re-use of all of the available frequencies within a cluster of 9-12 cells.

The performance of the handover process critically affects the user perception of system performance and integrity in terms of general maintenance of call quality, minimisation of dropped or lost calls, and temporary channel interruption (clicks) occurring during the handover itself.

Thus the planning and control C/I levels and handover processes are crucial to the overall performance of the system.

14.5.2 The measurement processes

Essential to the effective control of any system parameters such as interference levels and handover procedures is an effective feedback mechanism. This allows the system to gauge the effect of, and forming the basis for, decisions and control actions. The GSM system has been designed to support a range of powerful measurement processes which can be used to support system control mechanisms. These are received signal level, receive signal quality and mobile to base station distance.

There are also currently proposals for measurement of the relative timing difference between the current and neighbouring cells which would provide additional information on the relative distance of the mobile station from other neighbouring cells.

14.5.3 Cochannel interference control

14.5.3.1 *Planning*

The initial means of determining and controlling C/I ratios in the system is through the radio coverage planning i.e. base station siting, mast heights and antenna radiation patterns, frequency planning and maximum transmitter output powers.

Once a system has been installed and is in service, however, there are further mechanisms for improving C/I and thus improving service quality (or permitting increased spectrum efficiency through more intensive re-use of frequencies).

Two of these mechanisms of particular interest are discontinuous transmission (DTX) and transmitter power control.

14.5.3.2 *DTX*

The DTX function in GSM ensures that during speech silences no data is transmitted and the transmitter is not turned on for that period. In this way the co-channel interference generated by transmissions is restricted to the periods when conversation is active and there is a corresponding increase in the overall mean C/I ratio which is a function of the activity statistics of conversational speech.

Current software models and estimates indicate that the use of DTX may increase mean C/I levels by as much as 2 dB.

One consequence of the use of DTX is a reduction in the confidence level of the signal level and quality measurements. There will now be some TDMA frames where nothing was actually transmitted and a receiver cannot readily distinguish between a signal which was transmitted and poorly received or one which was not transmitted at all! This produces a further complication in the transmitter power control and handover algorithms in the system which must weight reported measurements as a function of DTX being used or not.

14.5.3.3 *Transmitter power control*
- The power control requirement
  An additional means of reducing the overall co-channel interference levels within the system is through the use of adaptive control of the transmitter

powers employed during a call or connection. The aim of power control is to ensure that transmissions are made at the minimum effective power necessary to maintain the required service quality, rather than all transmission taking place at the maximum power necessary for service at the cell edge. The mean transmission power employed may thus be reduced with a corresponding reduction in mean co-channel interference levels within the system.

- The GSM system features
  In the GSM system, the possibility of independently controlling the transmitter powers employed by the base station (downlink) and mobile station (uplink) is supported. An overall control range of 30 dB in 2 dB steps down from the maximum is possible and the dynamic control of the transmitter powers during a call is an operator optional feature. The ability of the mobile station to alter transmitted power under control of signalling from the base station is a mandatory feature which is also employed to preset the maximum power to be used within any particular cell.
  Again, models of the system operation indicate that the use of adaptive power control during calls may increase the mean C/I levels within the system by up to 2 dB.

- The Algorithms
  In order for the system to make decisions on the need or otherwise for power control action to take place, the measured and averaged (including weighting to allow for the use of DTX) values of received signal level and quality must be compared with some preset thresholds. In the GSM system there are two thresholds defined for averaged values of each measurement. The reasoning behind having two thresholds rather than one is to produce a 'deadband' type of control response where if the measured parameter lies in the window between the thresholds then no action is taken. This introduces stability into the process and inhibits 'hunting'.
  An additional stability measure is the definition of a maximum update period whereby a power modify command cannot follow another before a minimum timeout period has expired.
  Interestingly, although intuition might suggest that the decision to adjust transmitter power should be based on the received signal level at the other end of the link, it is in fact more pertinent to base power control decisions entirely on received signal quality. This is for the reasons suggested briefly above i.e. the aim of power control is to maintain the subjective quality of the connection at the minimum transmitter power and the latter provides a measure of that quality regardless of the overall signal level at the receiver (which may be dominated by co-channel interference).
  The problem which then arises is that comparison of received quality with decision thresholds can obviously give rise to a decision to increase or decrease transmitter power but cannot indicate by how much. The relationship between any change in received quality and a change in wanted signal level achieved through power control is a complex one which involves factors such as fading environment, vehicle speed, cochannel interference level, etc., which we cannot readily measure in an operational system.

Possible approaches for determining the required step size include using received signal level as an input parameter and tracing the history of the connection to extrapolate the quality versus level relationship. Such an approach is being investigated but first systems are likely to be based on a simple algorithm using a fixed step size and implementing an up/down incremental approach rather than a 'go to level x' system.

In order to guarantee the required minimum acceptable channel quality (which must take a higher priority than the need to minimise the transmitter power used) it is desirable to weight the power control algorithm in favour of increasing rather than decreasing power. This can be achieved by several methods including judicious choice of upper and lower decision thresholds and different hysteresis or update period values for increase/decrease decisions. In the simple initial up/down systems it may be achieved by having a larger increase step size than decrease.

Another factor which is currently being investigated is the overall stability of power control algorithms is a large scale cellular radio system. It is feasible to imagine a scenario whereby a decision is made to increase transmitter power because of unacceptable performance on a particular connection. This increase of transmitter power would also have the effect of increasing the interference levels being caused by that particular transmitter in the next co-channel cell. The resulting decrease in the channel quality for the user in that cell could also result in a transmitter power increase action which further increases the interference in the original cell, etc.

Thus it can be seen that there is a positive feedback mechanism within the system which may result in every channel in the system always tending to increase power until everyone is constantly operating at maximum power (thus eliminating the benefit of the whole complex process!). In reality, the effect of increasing power in one cell on interference in the co-channel cell is not a direct 1:1 relationship. The decorrelating effect of the 'random' use of particular channels in each cell, the distribution of call activity with time, the geographic distribution of subscribers within the cells, and other features like DTX during calls, will all tend to reduce the impact of the positive feedback mechanism and drive the system towards stability. This particular mechanism, however, is currently under detailed study and is one of the reasons why it has even been suggested that power control, particularly on the downlink, may not work!

14.5.4 Handover

14.5.4.1 *Introduction*
The handover process in a cellular radio system is used to pass a call from one base station to another as the mobile station concerned moves within the service area. In this way the system can maintain calls even though the user may be moving relatively rapidly through the areas (cells) served by each base station.

There are two possible approaches to making the handover decision. In the first, the handover decision is primarily based on the performance of the current channel falling or threatening to fall below an acceptable threshold. Handover is then attempted to an alternative base station which may be able to provide satisfactory service. In the second, handover is performed on the basis of potentially improving the quality of connection, even if the current one is deemed acceptable.

It should be appreciated that the handover algorithm employed within the system must be coupled to the corresponding power control algorithm if power control is being employed. It would be unacceptable that the system could enter into the handover process on the basis of a failing channel quality which could be maintained by simply increasing transmitter power. Conversely, it would be unacceptable to issue a power control command immediately prior to a handover action, etc. Also both power control and handover decisions are based on the same set of signal level and received quality measurements which must be weighted and/or discarded following an action (e.g. previous received signal level averages become invalid for the channel immediately following a transmitter power change, all measurement processes restart following a handover to a new BS). Thus each decision process must take account of actions arising from the other.

The subtle difference in the two handover decision approaches produce quite different requirements for the controlling algorithms and in particular lead to very different methods of coupling the handover and power control algorithms.

14.5.4.2   *The 'Minimum Acceptable Performance' method*

In this method the decision to begin the handover process is simply based on the averaged and weighted received signal level or received signal quality values for the connection falling below minimum (programmable) thresholds. In this event, the system will then attempt to identify the best candidate 'target' cell on the basis of the neighbouring cell received signal level reports received from the MS if it is expected that any of these can produce an improvement. If none of the candidates is likely to improve performance then the current connection is retained (as long as possible).

The decision as to whether to use received signal level or received signal quality as the comparison parameter is subject to similar arguments as described for power control. Ostensibly, the decision should be made on received signal quality since this is the 'real' measure of the prevailing channel performance. It must be appreciated, however, that the subsequent identification of a suitable handover candidate can only be made on the basis of relative received signal level measurements (we currently have no means of measuring received signal quality on neighbouring cells) which may not necessarily be a true reflection of the potential performance of each candidate connection.

The effect of this uncertainty may be reduced by the introduction of a neighbouring cells hysteresis (handover margin) parameter whereby handover will occur to a particular candidate if the potential received signal level from that candidate exceeds that from the existing cell by the margin value. Careful determination of these programmable parameters for every cell pair can be used to increase the probability of successfully improving performance by handing over, but at the possible expense of increasing the (low) probability of deciding not to handover after all when it may have been beneficial!

The action of this handover approach can be summarised thus:

"If the performance of the current channel is definitely unacceptable then attempt a handover to that BS which seems most likely to be acceptable".

The coupling between power control and handover algorithms in this approach is relatively simple. In this instance power control takes precedence over handover decisions, i.e. the handover comparison process need not be performed until the maximum permissible power for the cell is being used.

One consequence of this relatively simple type of handover algorithm (which is used extensively in existing systems) is that cell boundaries become 'smeared' since a connection with a given BS will always be maintained as long as the performance is acceptable. Thus users travelling from cell A into cell B may get some way into the new cell area before handover occurs, a similar process occurs in the other direction. This smearing of cell boundaries can make coverage prediction and frequency planning very difficult and can tend to result in increased cochannel interference levels due to MS continuing to transmit on a given channel until well outside the planned coverage area.

For very large rural cells this effect may be reduced by the use of the MS-BS distance measurement as an additional criterion in the handover decision process. The accuracy of measurement of the MS-BS distance is itself limited, however, and limits its use in smaller, urban cells.

14.5.4.3 *The 'Most Suitable Connection' method*

In this method, which is described in GSM as the *'power budget'* method, the handover process aims to hand a call over to any BS which can provide a better or equivalent service quality at the same or a reduced power level.

This is achieved by evaluating a value of the power budget, PBGT, for the existing connection and the neighbouring cells based on received signal level and the maximum transmitter powers permissible in each cell (or the maximum MS capability). The PBGT provides a measure of the difference in path-losses which could be accommodated between connections with the current and other potential BS. If this is found to exceed the handover margin (hysteresis value) for the neighbouring cells then there is a high probability that handing over to that target cell will result in increased received signal level or reduced transmitter powers being necessary.

The value of power budget calculations of this type is that it is possible to achieve a high probability of handovers to BS with improved received signal quality and there is little smearing of cell boundaries.

Another major advantage of this method is that it can be used to support the *'umbrella cell'* concept whereby some large cells (umbrella cells) are overlaid on the network of smaller cells in order to mop-up coverage area blackspots. With the previous method of handover decision an MS which was connected to a channel on the umbrella BS would not handover to the smaller cells until performance became unacceptable. The second method would result in handover to a smaller cell which could provide the required signal level, i.e. the desired result.

The primary disadvantage is that the whole process is heavily based on received signal level which, as we have seen, is not necessarily indicative of better service or channel quality. Again, the probability of hand over to a BS with better received signal level but poorer received signal quality can be reduced by careful choice of handover margin parameters but increasing handover margins will increase cell boundary smearing once more.

In the case of handover decisions being made on the PBGT basis, the coupling with the power control algorithm is relatively simple. In this case the handover algorithm takes precedence and the power control decision process takes place following a decision not to take handover action. The power control decision can still be based primarily on received signal quality rather than received signal level.

14.6   Conclusions

There is no disputing the fact that GSM is a very complex system but it is generally considered that this is the price that has to be paid to achieve a performance level nearing that of the fixed network and to achieve a subscriber capacity able to meet the predicted demand. It is expected however that the total volume of equipment needed to support a subscriber base of 10 million by 1999 will allow prices to remain competitive with current analogue systems. Furthermore the challenge will give a significant impetus to technology development in Europe and it is also likely to have substantial spinoff benefits to other radio and telecommunications services and products.

Acknowledgement

The author is in the fortunate position of being the leader of a highly motivated and talented team of engineers. Most of the ideas and some of the text in this paper come directly from them and the author would like to express his appreciation for the support they have given.

14.7   References

1.   NATVIG, J.E.: 'Speech coding in the pan-European digital mobile radio system', Speech Comms. Mgz., January, 1988

2.   MUROTA, K., and HIRADE, K.: 'GSM modulation for digital mobile radio telephones', IEEE Trans, COM-29, No.7., 1981

3.   ANDERSON, J.B., AULIN, T., and SUNBERG, C.I.: 'Digital phase modulation' (Plenum Press, 1986)

4.   BALSTON, D.M.: 'Pan-European cellular radio', IEE Elec. & Comms. J., 1, 1989

Chapter 15

Future personal telecommunications

Heinz Ochsner

15.1 Introduction

In order to be able to discuss Future Personal Telecommunications (FPTC) we need to understand the term personal telecommunications as such. Many views about personal telecommunications exist. The way we look at it here is as a (technical) communication means which is available to any person anytime and everywhere, i.e. universally. Availability is understood in terms of being able to contact somebody, as well as being able to be contacted by someone else.

Obviously this view of personal telecommunications requires that the user's communication identity, typically the directory number, is always with the person. Rather than numbering subscriber loops as in the wired telephone networks, we speak here of *personal numbering* or Identity Number (ID). Such a communication ID may be in a communication terminal carried by the user, the ID may however be contained in a module separate from the terminal, e.g. in a smart card. While we will concentrate on the former aspect, i.e. on a small terminal carried by the person, the aspect of separate identification modules should not be forgotten. It should be noted the concept of personal numbering is independent from the way the user identification is implemented. Personal numbering is a network feature rather a user (or user terminal) feature.

While today we usually think of a telephone when talking about personal telecommunications we should not forget that in future this service should satisfy *all* communication needs.

A communication terminal which is carried around needs to be "pocketable". The limits for a pocketable terminal are broadly accepted to be 200g in weight and 200cm^3 in volume. Furthermore we expect reasonable battery life-time. These and other requirements are discussed in section 15.4.

In summary: personal telecommunications offers all communication services *anytime* and *everywhere* to *everybody* by the use of a single identity and pocketable communication terminal[1].

---

1 For simplicity, we include a smart card in the expression "terminal".

15.2 Personal telecommunications today

In today's telecommunication world we are faced ourselves with a wide variety of internationally (or even nationally) incompatible systems for different services. The two aspects: *service* and *international compatibility,* are treated separately.

15.2.1 The service aspect

Currently, i.e. 1990, we can distinguish six different groups of services (see Fig.15.1). While *wired* and *cordless telephony, paging,* and *cellular radio* are services offered by public networks (i.e. available to everybody), *private mobile radio* or *airfone services* are non-public. Note that all these six areas of service may include non-voice services as well.

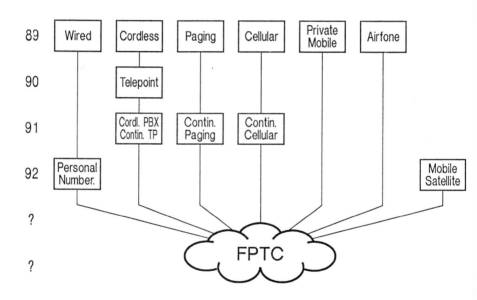

Fig. 15.1   Evolution towards personal telecommunications

As Fig.15.1 shows, new services will be available during 1990/91, typically *Telepoint* or *wireless PBX*. Also in 1991 we expect cordless telephony (including Telepoint), paging and cellular radio to be available as European networks. These systems, known as DECT, ERMES, and GSM will be treated in section 15.3.

From now on we will concentrate on the services offered by today's public telecommunication networks.

Fig.15.2 shows a comparison of the typical properties of personal telecommunications with those offered by today's public networks. Since Telepoint is expected to be introduced very soon, the column "Cordless" includes the comparison with Telepoint as well.

A black bullet in the table says that the point is fully supported, while a circle denotes only limited support. Obviously, we expect the personal telecommunicator to fully support all points.

	Wired	Cord-less	Cell-ular	Paging	FPTC
Voice services	●	●	●		●
Non-voice services	●	○	○		●
Message services	●			●	●
Complete local coverage	○	●		○	●
National networks	●	○	●	●	●
International networks	●		●	●	●
International roaming			○	○	●
Personal numbering		○	○	○	●
Pocketable terminal		●	○	●	●

Fig. 15.2 Today's public services and personal telecommunications

Notes for diagram:

1. *Complete local coverage* means availability of the service everywhere within small areas, i.e. within buildings in every room or corridor, etc.
2. *International networks* may well exist, while *international roaming* means that the personal identity can be used internationally. Some form of *personal numbering* is required for international roaming.

15.2.1.1 *Wired networks*
No personal numbering exists in wired telephony, the directory number denotes a subscriber loop, or an outlet. Therefore, no roaming is possible. Some functions which go towards personal numbering, such as call diversion, "follow me", exist in private networks only. They are expected in public ISDN.

Local Area Networks (LAN) go one step further: The identity resides in the terminal (or modem).

One major drawback of wires is their lack of continuous coverage. Therefore wired networks are not available anytime everywhere.

15.2.1.2 *Cordless telephony and telepoints*
Cordless telephones (CT) solve the wired telephone's coverage problem in a small area, but a cordless phone is still an ordinary telephone; however, recent CT products are quite close to being pocketable in the sense of weight and size.

Telepoints will extend some applications of cordless telephones and give access to national networks. It is also interesting to note that public telephone operators today consider replacing the local loop by a radio link using technology similar to cordless telephones and Telepoints.

Telepoints will provide a kind of limited personal numbering, since the terminal with the ID usually travels with the person.

15.2.1.3 *Cellular radio*
Cellular radio today is to some extent the classical car telephone. While nice hand-portable terminals are now on the market for various cellular systems, really pocketable terminals do not yet exist.

The cellular networks usually do not provide complete coverage within buildings. Since they have been designed for car telephony they are often unsuitable to cope with the traffic requirements and propagation properties within buildings.

As with Telepoints, cellular radio provides a limited personal numbering.

15.2.1.4 *Wide area paging*
Wide area paging networks provide only very limited services, in particular they convey short messages in one direction only. However, except for the range of services offered, paging is very close to personal telecommunications.

15.2.2 The compatibility aspect

15.2.2.1 *Wired telephony*
Despite the availability of a worldwide telephone network there is no reasonable compatibility of the subscriber loop even throughout Europe. Because the national networks have evolved individually and under the national authorities' monopolies, technical parameters as fundamental as line impedances, loudness ratings, or dialling signals differ from country to country. Some compatibility can be expected throughout the world with ISDN.

15.2.2.2 *Cordless telephony*
Four different and incompatible systems exist in Europe, (Fig. 15.3). An early approach to standardise cordless telephony is CEPT CT1 (CEPT Rec. T/R 24-03). Today CEPT CT1 exists in two versions, one with 40 channels, the other with 80. The CEPT CT1/40 version exists in Norway, Sweden, Denmark, Holland, Belgium, Germany, Austria, Switzerland, and Italy. New equipment for Germany and Switzerland has to conform with the 80 channel version (on different frequencies). Other countries are expected to change to this version as well.

There is no interoperability among products of different manufacturers within all four systems. The specification provides mutual co-existence only. This is adequate for classic cordless telephony. Telepoint services, however, require a common air interface specification for interoperability among different manufacturers' products. Such specifications are currently being developed in the UK for CT2 (CAI - Common Air Interface) and in Germany for CEPT CT1/80 (ELSE - Einheitliche Luft-Schnittstelle).

It is noted that the more advanced systems UK CT2 and CEPT CT1 use dynamic channel allocation, i.e. all terminals (or base stations) choose a free channel among all the available channels. Dynamic channel allocation is a requirement to cope with the unpredictable traffic densities and propagation characteristics within buildings.

	CT1	CT2	CT (F)	CT1
Frequency band (MHz)	1.6/47	860	26/49	915/960 (885/930)
Number of channels	8	40	15 allocated 10 in use	40 (80)
Allocation	fixed	dynamic	fixed	dynamic
Technology and access	analog FDMA/FDD	digital FDMA/TDD	analog FDMA/FDD	analog FDMA/FDD

Fig. 15.3   European cordless telephone systems

15.2.2.3 *Cellular radio*

There are four major modern systems (i.e. developed in the seventies) in use throughout Europe (Fig. 15.4). Radiocom 2000, being the oldest one of these systems, is not really cellular since it does not provide handover. Radiocom 2000, however, combines car telephony with a dispatcher service. Note also, that in most countries older generations of car telephone systems are still in use.

The main difference between the systems are the frequency bands, channel separation, and multiplexing of traffic and signalling data. All systems use special call set-up channels for mobile terminated (MT) calls. Mobile originated (MO) calls are set-up in the NMT system using any free speech channel. Radiocom 2000 and C450 provide off-air call set-up (MO), i.e. the speech channel is assigned only after the set-up in the network is completed.

Although technically possible in all systems, international roaming is offered by the NMT operators only.

Note that all systems have a fixed assignment of some of the channels to each cell. The dependence of the signalling protocols on this fixed channel allocation makes cellular systems inappropriate for use within buildings.

	TACS	NMT	Radiocom 2000	C450
Frequency band (MHz)	900	450 900	200 400	450
Number of channels	1000	180 1000	192 256	222
MO call set-up chan.	dedicated	any free speech ch.	dedicated (off-air)	dedicated (off-air)
MT call set-up chan.	dedicated	dedicated	dedicated	dedicated
European countries	GB, I, A	N, S, DK, IS, SF, CH, NL	F	D, P
International roaming	no	yes	no	no

Fig. 15. 4   European cellular radio systems

15.2.2.4 *Wide area paging*

There are three main parameters which are changeable in wide area paging systems: Frequency, bit-rate and coding. Unsurprisingly, about all possible combinations exist in Europe as set out in Fig. 15.5. A first step towards a European system including roaming is Europage.

		Frequency MHz	Code	Bitrate bit/s
▨	United Kingdom	138/150	POCSAG	512/1200
▮▮	France	450	POCSAG	512
▬	Germany	450	POCSAG	512
▮▮	Italy	450	Golay	512
✚	Switzerland	150	POCSAG	512
✚	Denmark	450	POCSAG	512
✚	Sweden	150	POCSAG	512
◉	Europage (GB,F,D,E)	450	POCSAG	1200

Fig. 15.5   European wide-area paging systems

15.3 The next generation digital European systems

The European Telecommunications Standards Institute (ETSI) is currently standardising three radio communication systems which are comparable to personal telecommunications: the pan-European cellular system GSM, the Digital European Cordless Telecommunications (DECT), and the European Radio Messaging System (ERMES).

15.3.1 GSM

GSM, named after the originally CEPT now ETSI technical committe *Groupe Spécial Mobile*, is the pan-European car telephone system. It was originally designed and optimised for use in fast moving vehicles. Nevertheless, hand-portable terminals are expected on the market.

GSM uses a Time Division Multiple Access (TDMA) with Frequency Division Duplex (FDD) on a total of 124 carrier pairs in the 900 MHz band. One carrier may bear 8 speech conversations with speech encoded digitally with rate 13 kbit/s. The frequency carriers are pre-assigned to the cell cites, i.e. GSM uses fixed channel allocation. This is the main reason why GSM still is not suited for in-house use, where the service area for one base station is predominantly very small with unpredictable propagation and traffic density environments.

A signalling system had to be created for GSM to allow efficient routing of calls to roaming users throughout Europe.

15.3.2 DECT

The prime aim of digital European cordless telecommunications is to cover the typically short range applications of cordless private networks (PBX and LAN), as well as Telepoint and Wireless Local Loop. The design of the DECTstandard is led by the need of pocketable terminals and in-house operation.

The radio access used in the DECT system is Time Division Multiple Access (TDMA) with Time Division Duplex (TDD) on 10 to 12 carriers around 1.9 GHz. One carrier will bear 12 duplex conversations with speech encoded using 32 kbit/s ADPCM. To cope with the traffic and propagation environment within buildings, dynamic channel allocation is used. Thus, in each cell all channels available channels may be used.

15.3.3 ERMES

The European radio messaging network is still a paging system, i.e. it will still offer only limited service capabilities. ERMES will use two carriers around 170 MHz and a completely newly designed code.

The main advantage over DECT or GSM will be the very small size of the terminals (which are only receivers) and the expected low prices compared to the speech transceivers used in DECT or GSM.

Fig.15.6 shows the different application areas accessing the radio medium which we finally expect to be covered (perhaps with the exception of mobile satellite) by personal telecommunications. Areas are shown in terms of service area (per base station) versus traffic density. Note that the Telepoint and Wireless Local Loop applications could potentially be covered by both DECT and GSM while for the other areas only one of the technologies is applicable.

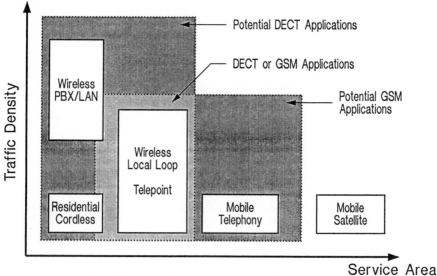

Fig. 15.6   Application areas of radio access

As was done in Fig.15.2, the planned systems GSM, DECT, and ERMES are compared with the capabilities that are deemed to exist for FPTC and are shown in Fig. 15.7.

	GSM	DECT	ERMES	FPTC
Voice services	●	●		●
Non-voice services	●	●		●
Message services	●	●	●	●
Complete local coverage	○	●	○	●
National networks	●	●	●	●
International networks	●	●	●	●
International roaming	●	●	●	●
Personal numbering	○	○	○	●
Pocketable terminal	○	●	●	●

Fig. 15.7   Capabilities of GSM, DECT and ERMES compared to FPTC

All three systems show more black bullets than the comparable systems of today. Therefore, all three approaches are important steps towards the personal communicator. DECT has only one circle left.

The user identity of a GSM subscriber will be contained in a Subscriber Identity Module (SIM). The SIM may be a exchangeable smart card. But even then the personal numbering is bound to GSM. It is currently under discussion whether a similar approach, perhaps using an identical system, could be used in DECT. This would then extend personal numbering to both GSM and DECT together.

15.4 User requirements in future systems

Let's recall the user expectations of a personal communications service:

> Personal telecommunications offers all communication services *anytime* and *everywhere* to *everybody* by the use of a single identity and pocketable communication terminals.

In addition to the service expectations, the use of the service should be cheap and safe. Therefore, we can put some requirements concerning system capacity (or traffic density), service coverage, battery autonomy, terminal size and weight, terminal and service cost, as well as human and environmental safety.

15.4.1 Capacity or traffic density

To guarantee access to personal telecommunications to everybody anytime the system must provide enough capacity, in particular enough radio channels. A recent publication [1] states a total system requirement of at least 200 duplex channels just to satisfy the need of telephony. This number does not yet take into account high traffic density situations or the need of non-voice (possibly high data-rate) applications. Up to 200 MHz of radio spectrum has been claimed for within the RACE and CCIR IWP 8/13 frame work (see section 15.5 below).

15.4.2 Service coverage

The personal telecommunications is expected everywhere. The concept must allow availability of the service inside and outside buildings, in high traffic density areas, as well in regions with relatively low traffic density. This latter case is solved in the classical cellular systems, including GSM, by large cells. However, in large cells the additional problems of long range communication, e.g. increased output power, and time dispersion, which may need equalisation, occur. Increased power consumption, increased circuit complexity, and finally increased terminal cost are the result. On the other hand, small cells even in low density areas increase the infrastructure, and therefore the service cost. The problem of covering all service areas in a cost effective manner has yet to be solved.

15.4.3   Battery autonomy

It is very difficult to assess the requirement for battery autonomy of the personal communicator. Many people expect that they can use their equipment several days without changing a battery or recharging the unit. Service availability anytime requires that there shall never be a need to switch off the unit just to save power. It is also noted that ecological considerations discourage the use of non-rechargeable batteries, e.g. of the mercury or lithium type.

15.4.4   Terminal weight and size

It has already been mentioned earlier that a current view is that the size of a terminal should be less than 200 cm³ whereas the weight should be below 200 g. It should also be borne in mind that arbitrarily small and light equipment becomes impractical to use as well. Pocketable equipment should be aimed for rather than the wrist-watch type of communicator.

15.4.5   Cost

Equipment and service has to be affordable for everybody !

15.4.6   Human and environmental safety

Effects of microwave radiation on living organisms, in particular humans are studied throughout the world. In view of the ever increasing frequency bands used these results have to be considered in the design of personal telecommunications. As already mentioned, environmental aspects of the use of non-rechargeable batteries need to be considered as well.

15.5   World-wide activities in personal telecommunications

Some of the world-wide activities in personal telecommunications need to be mentioned:

RACE   The *Research into Advanced Communications in Europe* programme number 1043 includes activities towards personal communications.

CCIR   The CCIR Interim Working Party IWP 8/13 studies the *Future Public Land Mobile Telecommunication System (FPLMTS)*. The IWP's view today is however closer to personal communications than to the classical mobile communications.

CCITT   Several CCITT Study Groups work on problems of personal numbering, signalling requirements, etc. to cope with the network requirements of personal telecommunications.

BELLCORE The US Bell Communications Research Laboratories are studying the wireless local loop.

DOC Canada's Department of Communications is conducting a programme on low power wireless communications in the workplace.

ESCORT This is the Japanese enhanced cordless telephone programme.

IEEE The IEEE 802.4L standard will cover spread spectrum short range data links using the ISM band.

15.6 Summary

We have given an overview on some aspects of personal telecommunications as they may be expected in the medium term future. Starting with an idealistic view of Personal Telecommunications, systems which are currently operating and systems which are currently being standardised have been compared. It has been shown that the current situation exhibits a variety of internationally incompatible systems for different services. This problem will only partly be solved with the new European systems GSM, DECT, and ERMES which still offer different services and cover different service areas. In a second part, the user requirements for the future personal telecommunications have been briefly reviewed, and it has been expressed that this is a service which has to be available for *everybody, everywhere, anytime.*

15.7 References

1. MOTLEY, A. J., and BUTTON, J. E.: 'Small cell engineering for personal communications', 2nd IEE national conference on telecommunications, York, 2-5 April, 1989

2. OCHSNER, H.: 'GSM - overview over the radio subsystem', Digital cellular radio conference DRCR, Hagen (FRG), 1988

3. TUTTLEBEE, W. H. W. (Ed): 'Cordless telecommunications in Europe' (Springer Verlag, London, 1990)

4. Van DIEPENBECK, C. T. W.: 'DECT, a general overview', Fourth Nordic seminar on digital mobile radio communications, DMR IV, Oslo, 1990

5. BUD, A.: 'System and networks aspects of DECT', Fourth Nordic seminar on digital mobile radio communications, DMR IV, Oslo, 1990

6. OCHSNER, H.: 'Radio aspects of DECT', Fourth Nordic seminar on digital mobile radio communications, DMR IV, Oslo, 1990

Index